Lecture Notes in Statistics **168**

Edited by P. Bickel, P. Diggle, S. Fienberg, K. Krickeberg, I. Olkin, N. Wermuth, and S. Zeger

Springer
New York
Berlin
Heidelberg
Hong Kong
London
Milan
Paris
Tokyo

Susanne Rässler

Statistical Matching

A Frequentist Theory, Practical Applications, and Alternative Bayesian Approaches

Springer

Susanne Rässler
Institute of Statistic and Econometrics
University of Erlangen-Nürnberg
Lange Gasse 20
D-90403 Nürnberg
Germany
susanne.raessler@wiso.uni-erlangen.de

Library of Congress Cataloging-in-Publication Data
Rässler, Susanne.
Statistical matching : a frequentist theory, practical applications, and alternative Bayesian approaches / Susanne Rässler.
p. cm.— (Lecture notes in statistics ; 168)
Includes bibliographical references and index.
ISBN 0-387-95516-X (softcover : alk. paper)
1. Statistical matching. I. Title. II. Lecture notes in statistics (Springer-Verlag) ; v. 168.
QA276.6 .R37 2002
519.5—dc21 2002070554

ISBN 0-387-95516-X Printed on acid-free paper.

Printed in the United States of America.

9 8 7 6 5 4 3 2 1 SPIN 10881911

Typesetting: Pages created by the author using a Springer $\TeX$ macro package.

www.springer-ny.com

Springer-Verlag New York Berlin Heidelberg
A member of BertelsmannSpringer Science+Business Media GmbH

To my fathers – by birth, by law, and by academia

Preface

Following a long but now fading German tradition, usually academic researchers are requested to write a habilitation thesis in order to qualify before applying for a full professorship anywhere in Germany. This book is the result of some years of research about statistical matching and a revised version of my habilitation thesis. The text is intended to be a reference book for applied statisticians and methodology-oriented researchers seeking comprehensive information concerning statistical matching. It covers a wide range of different aspects of what in Europe is typically called data fusion. The necessary frequentist and Bayesian theories are presented in Chapters 2 and 4. Practical applications are described in Chapter 3. Chapter 5 finally provides an evaluation of the alternative Bayesian approaches with real-world media data. I hope the reader finds the material presented here to be a stimulating and useful contribution to the theory and practice of statistically matching data from different sources.

After I finished my doctorate degree with a work concerning survey sampling I was "pushed on" by the head of our chair, Ingo Klein, to write this second thesis. He was the first to arouse my interest in statistical matching and I am very grateful in his belief in my ability to produce some useful contributions. My friend and former colleague, Karlheinz Fleischer, proved to be an excellent partner and together we derived the frequentist approach presented herein. We entered into some lively and engaging discussions with our German marketing research company, GfK AG, about the assumption of conditional independence which is implicitly made when different surveys are matched. I am much obliged to them for the open discussions we still have and for providing me with data. A special thank you has to be said to Friedrich Wendt who gave me access to all his private literature and procedures. He was the first, in Germany if not everywhere, to merge data from different sources.

When I contacted researchers working on the problem of nonresponse I found great interest in the topic of statistical matching that can be viewed as a special case of nonresponse. Following an idea proposed by Rubin in his 1987 multiple imputation book, I became involved with multiple imputation. The excellent 1997 book by Joseph L. Schafer about the analysis of incomplete data was my introduction to the world of multiple imputation. Joe has always been extremely helpful and friendly in explaining the correct way how to apply multiple imputation techniques. Finally, I was lucky to meet Donald B. Rubin himself whom I regard to be my remote teacher. The more I know the more fascinated I am by his multiple imputation techniques. Thus, it was quite obvious for me to apply such alternative Bayesian approaches to the matching task and derive a suitable multiple imputation technique. Clearly, there is more work to be done and I could have waited for improved results from ongoing research. However, there was no comprehensive book about statistical matching available and, among others, Joe as well as Vasja Vehovar encouraged me to look for a publisher. I was fortunate to find an outstanding publisher with Springer-Verlag. John Kimmel's cordiality and the friendliness of his editors made it most enjoyable to work with them; thanks also go to the reviewers for their very helpful suggestions and detailed comments as well as to Daniel Gossel and the Springer copyeditors for their corrections of my English.

Finally, I am very grateful for the cooperative atmosphere at our Chair of Statistics and Econometrics, and, moreover, for all the help and support I got from my beloved family and my friends, especially my husband, Hendrik, who was so patient with me. Besides all our businesses, he became an excellent cook and always encouraged me to keep on and finish the habilitation and, now, this book. Unfortunately my father-in-law did not live to see the end of my habilitation. He had always been proud of my doing what he himself discarded in the late 1950s when he and his family left the former Eastern German Republic for political reasons.

Mittelehrenbach, June 2002 Susanne Rässler

Contents

List of Figures

List of Tables

1
Introduction

I might describe statistical matching as imputation out of desperation and against our better judgment. [...] Heroic appeals to conditional independence then produce imputed microdatasets from which one can supposedly draw inference about the impact of Y if X is changed. Further discussion of that approach [...] is unnecessary given the thorough discrediting by Rodgers (1984).

D.R. Judkins (1998)

1.1 Statistical Matching – Problems and Perspectives

It seems that statistical matching splits the field of statistics in two. Statistical matching is blamed and repudiated by sceptical theoretical and practical statisticians about the power of matching techniques. This is reported, e.g., by Moriarity and Scheuren (2001), Judkins (1998), Gabler (1997), Bennike (1987), Rodgers (1984), Woodbury (1983), and Sims (1972a and b). On the other hand, famous statistical offices such as Statistics Canada as well as market research companies especially in Europe have done or are still doing statistical matching which in Europe is typically called data fusion. However, from time to time there are reports published stating that data from different sources have been matched successfully. Positive experiences with statistical matching have been published in a wide variety of journals or as internal reports or working papers, e.g., by Aluja-Banet and Thio (2001),

Wendt (1976, 1986, 2000), Kovacevic and Liu (1994), Liu and Kovacevic (1996, 1997, 1998), Roberts (1994), Baker (1990), Baker et al. (1989), Antoine (1987), Antoine and Santini (1987), Scheler and Wiegand (1987), Wiegand (1986), Okner (1974), Ruggles and Ruggles (1974), Ruggles et al. (1977), and Okner (1972a and b).

To start an introductory discussion about the advantages and disadvantages of statistical matching, we first give a short insight into it. Throughout the world today we find synonyms used to describe the statistical matching process including data fusion, data merging or data matching, mass imputation, microsimulation modeling, and file concatenation. During the 1980s a discussion dragged on about a suitable and clarifying denotation. Finally, it was decided to leave "injection", "imputation", "attribution", "data ascription", "data integration", or "data connection" to be reserved for more general use; see, e.g., Antoine (1987). In European marketing literature and practice the most commonly used term today is "data fusion". Not to bother the reader with the messy terminology we mostly write "statistical matching" throughout the book, and occasionally use "data fusion" synonymously.

Historically, statistical matching was motivated by the interest in people's consuming behavior especially to improve media targeting. Empirical studies concerning the association between individual television viewing and purchasing behavior, for instance, can be difficult in the majority of cases. The ideal medium would be a very large consumer panel, where each individual's purchasing and television viewing behavior were both being measured over a long time period. However, running such a large so-called single source panel often is impractical or its costs prohibitively high. Furthermore, a high percentage of nonrespondents or poor quality of data is to be expected; see, e.g., O'Brien (1991). A test performed by the GfK Nuremberg in 1995, the largest German market research institute, showed the consequences if purchasing behavior data are collected in addition to TV viewing behavior data as leading to an extremely high rate of panel mortality, to changes in TV viewing behavior, and, finally, to remaining households not being representative for the population of interest; see Darkow (1996) or Wildner (1994).

Thus an attractive alternative is to make use of a statistical matching technique to link, for example, the viewing information available from a television measurement panel with the purchasing data available from an existing large market tracking panel by means of some variables common to both data files. Figure 1.1 illustrates the principle of statistical matching using a very simplified example.

As stated above, basically, statistical matching was initiated by media targeting demands. In particular, applications that combine product usage and TV viewing behavior have long been implemented as enhanced media planning tools. Variables that have not been jointly observed are imputed on an individual level creating a large and complete database. Without loss

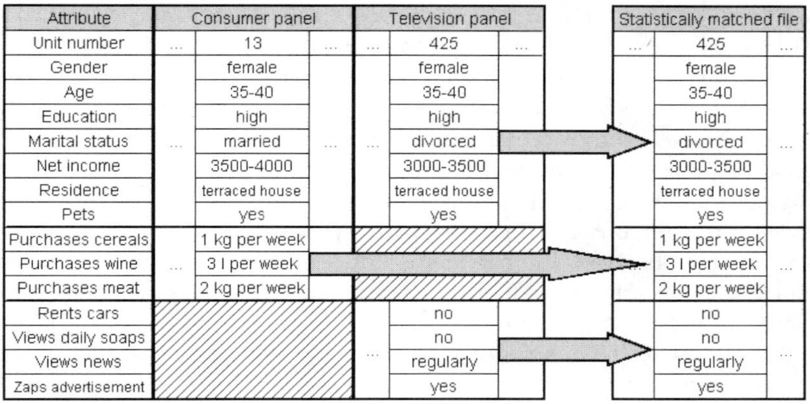

Figure 1.1. Illustration of statistical matching

of generality we can assume that the overlap (i.e., the number of identical individuals in both samples) is negligible, if there is any at all. Therefore the merging usually is done by means of a particular distance function between the units in both files, which are often called donor and recipient units, respectively. Considering Figure 1.1 we would, for example, search for a donor unit (element of the consumer panel) which is most similar (i.e., a so-called nearest neighbor) to the recipient unit (element of the television measurement panel) compared by their common variables. Finally, the consuming information of the donor is added to the recipient unit leading to a new and complete but artificial statistically matched file. Once the data are matched the analysis proceeds as if the artificial statistically matched file were a real sample representative of the true population of interest.

In the last decades papers have been published showing that traditional techniques of statistical matching establish so-called **conditional independence**; see especially Rodgers (1984). Under conditional independence the variables never jointly observed are independent given the variables observed in both files after the matching has been done. Referring to Figure 1.1 in the artificial statistically matched file we find the TV viewing and the purchasing behavior being more or less (conditionally) independent given the demographic and socioeconomic information. So the gain of statistical matching is known a priori. Theoretically speaking, there are alternative strategies for many types of analysis. For example, assuming a normal model and conditional independence, the unknown association between the variables never jointly observed can be estimated from the two separate files. Practically speaking, the creation of a file of microdata records with measures of all variables is attractive because of the flexibility it offers the analyst. This is particularly true if the data are to be used by many analysts for many different purposes as is often the case in market research, or, finally, if the data are to become a public resource.

Thus, what is the controversy about? From an information-theoretic point of view it is easy to accept that the association of variables never jointly observed may not be estimated from the observed data. Rubin (1974) "...presents a framework for organizing and understanding the problems of estimating parameters of a multivariate data set which contains blocks of missing observations." He shows that whenever two variables are never jointly observed, the parameters of conditional association between them given the other variables are inestimable by means of likelihood inference. Nevertheless, many matching techniques mainly based on nearest neighbor matches have been applied over the years. But these traditional approaches to statistical matching establish (conditional) independence between the variables never jointly observed given the common variables. Hence critical voices argue that any statistical matching appears to be unnecessary because the outcome is already well known. Moreover, conditional independence is produced for the variables not jointly observed although they may be conditionally dependent in reality. The critics are right so far.

However, usually the analyst is not only interested in the conditional association, e.g., of viewing a particular channel and purchasing cereals given all common variables, but is focused on the unconditional association to find the best advertising time and channel to publicize one's business. Often analysts' interests are directed to the unconditional association or to the conditional association of the variables never jointly observed given some few common variables such as gender, age, or type of household. Clearly, under a specific data model, for instance, assuming a multivariate joint normal distribution with a common covariance matrix we may calculate the unconditional association given conditional independence for some variables or any other conditional association of interest. Also the range of the possible unconditional associations of the variables not jointly observed due to the joint model can be derived theoretically. However, these calculations depend on the data model assumed becoming more complex the higher the dimension of the data is. We have reached the root of the matter. How can we seriously handle hundreds of variables on purchasing and viewing information given a hundred common variables? Moreover, these variables typically have skewed distributions, others follow semicontinuous distributions, or are of categorical type. For the moment it seems impossible to find a sufficiently suitable joint data model for the statistical matching task from which to (theoretically) derive possible unconditional associations. Therefore we have ended up with heuristical approaches allowing at least the matching of large data sets from which any inference based on any individual analysis may be drawn. Advocates of statistical matching thus argue that if the common variables are (carefully) chosen in a way that establishes more or less conditional independence among the variables not jointly observed given these common variables, then inference about the actually unobserved association is valid. Unfortunately this implies another untestable assumption.

Hence we end up with the actual situation that offices and institutes are regularly matching data sets still using approaches leading to conditional independence and, on the other hand, people are disclaiming statistical matching.

In the meantime computational power has increased dramatically, and modern techniques such as Markov chain Monte Carlo methods have become more and more familiar and available allowing the solution of quite difficult estimation tasks. Also people are becoming more and more familiar with the concept of multiple imputation proposed by Rubin (1987) which is addressed to problems due to nonresponse and missing data in general. Roughly speaking, with multiple imputation the missing data are imputed more than once, and these imputations usually are based on some distributional assumptions. Multiple imputation is a general concept proposed to display sensitivity of any analysis because there is uncertainty about the missing data and, thus, about the values to be imputed. Although these methods have become more and more widespread recently and some software routines are now available, the application of multiple imputation to the statistical matching task is not self-evident due to the inestimability of certain parameters. Whenever the underlying model cannot be completely specified by the data, problems of identification occur. In the matching case we usually find more than one feasible association to describe the joint distribution of the variables never jointly observed. In other words, we cannot guarantee that these random variables could not have followed different joint distributions with different sets of association parameters. We call this the identification problem which is inherent in statistical matching.

First approaches to realize multiple imputation in the context of statistical matching were published by Rubin (1986) using an implicit model and by Rubin (1987), pp. 186–188, proposing an explicit model. The first, and as far as we know only, experiment to apply a multiple imputation procedure to a real statistical matching task was published by Kamakura and Wedel (1997), yet their approach was restricted to the estimation of bivariate cell frequencies. Hereinafter we propose some model-based approaches tackling the statistical matching task by more flexible techniques. By means of the Bayesian framework we make prior information usable. Moreover, multiple imputations will help us to reflect the identification problems that are inherent in statistical matching. We discuss the frequentist properties of the traditional approaches first and then give a deeper insight into those procedures applied in practice. Afterwards we introduce the concept and principle of multiple imputation. Existing procedures, which are successfully used in the context of missing data due to nonresponse, are reviewed and discussed in the light of statistical matching. Finally, a multiple imputation procedure especially suited for the matching task is proposed. Simulation studies are performed to illustrate the procedures under discussion and, last but not least, the techniques suggested herein are applied to some real media data to show their practical importance.

By the end of the book we have pointed out that efficient statistical matching is possible by means of suitable modeling and can even be improved by introducing prior information. We finally conclude that in our opinion statistical matching may be worth the effort.

1.2 Record Linkage Versus Statistical Matching

Rodgers (1984) states that much of the interest in statistical matching procedures arises from the analogy of statistical matching with exact matching which is also known as record linkage. Historically their roots are clearly related. With record linkage the two samples contain identical individuals that have to be found for the match according to the individual's name, address, or social security number. Therefore it is often necessary to deal with problems of errors in these link variables. Additional information such as age or gender may be used to increase the probability that merged records apply to the same individual, and to find the most probable match when unique identifiers disagree because of recording or reporting errors.[1] However, according to Rodgers (1984), "It is a relatively small step, computationally, from such procedures for exact matching of identical individuals to statistical matching of similar individuals. A small step for the computer is in this case a giant step for the statistician – a step that should only be taken with full awareness of the importance of the implicit assumptions and the potential consequences of the incorrectness of those assumptions."

With statistical matching the number of identical individuals in both samples is at least small if not zero. Therefore it is only possible to match "similar" individuals and, thus, the implicit assumption of conditional independence may arise.

Since this book tries to figure out all possible aspects of statistical matching we do not refer to record linkage again.

1.3 Statistical Matching as Nonresponse Phenomenon

Discussion of the matching task in the light of survey nonresponse is the best way to proceed. Nonresponse in questionnaires that leads to missing

[1]It seems that record linkage is still of greater interest to researchers than statistical matching. The majority of papers, e.g., presented at the Symposium 99: Combining Data from Different Sources, Statistics Canada, Ottawa, May 4–7, 1999, dealt with problems of record linkage; only four papers were dedicated to statistical matching.

data either on some or, at the worst, all items of the questionnaire has long been addressed by researchers and practitioners. Rubin (1987) and Little and Rubin (1987) once classified the nonresponse phenomenon according to the probability of response yielding the following three cases.

- The missing data are said to be missing completely at random (MCAR) if the nonresponse process is independent of both unobserved and observed data.
- If, conditional on the observed data, the nonresponse process is independent only of the unobserved data then the data are missing at random (MAR). This is the case, for example, if the probability of answering the income question varies according to the age of the respondents, whereas the age is observed, but not the income itself.
- A nonresponse process that is neither MCAR nor MAR is called nonrandom or missing not at random (MNAR) (i.e., the probability of a variable being observed depends on the variable itself). Missing income data, for example, are MNAR, if the probability that income is not recorded is higher, when the income is higher.

In the context of likelihood-based inference and when the parameters describing the measurement process are functionally independent of the parameter describing the nonresponse process, MCAR and MAR are said to be ignorable; otherwise we call it nonignorable missingness. Of course, the nonignorable case is the hardest to deal with analytically.

However, concerning statistical matching we agree with Kamakura and Wedel (1997) and Gelman et al. (1998) that the missing data are, at least, missing at random (MAR). In particular, this missing data mechanism is not stochastic as in many other missing data problems but deterministic, because it is induced by the study design of the two samples (i.e., all of the missingness is due to unasked questions). Therefore, this particular missing data mechanism is treated as ignorable corresponding to a very unique missingness pattern, because some variables are never jointly observed for all units of the files as illustrated by Figure 1.2.

The problem of nonresponse is increasingly being addressed in the literature. A wide range of techniques to fill in plausible values when a variable is not observed and its value is missing have been developed and improved. In the meantime modern imputation techniques are more frequently based on multiple imputations and Bayesian modeling. However, when variables are never jointly observed the problem of inestimability of parameters concerning the unobserved joint association arises which may be solved by using informative prior distributions. Rubin (1987), p. 187, states that "within the multiple-imputation framework it is not necessary to assume conditional independence or any other specific choice for the parameters of conditional association, because each set of imputations can be made

Consumer panel

Unit no.	Gender	Age	Education	...	Purchasing information about: cereals	wine	...	Viewing information about: daily soaps	news	...
1	female	40-45	low	...	1 kg per week	1 l per week	...			
2	male	30-35	high	...	none	2 l per week	...		missing data	
...	...	...	...	...	...	...	...			

Television panel

Unit no.	Gender	Age	Education	...	Purchasing information about: cereals	wine	...	Viewing information about: daily soaps	news	...
1	female	>60	low	...				regularly	no	...
2	female	30-35	high	...		missing data		no	regularly	...
...	...	...	...	...				...	...	...

Figure 1.2. Particular missing data pattern induced by the statistical matching task

for a different choice of parameters of conditional association." Since informative prior distributions are usually avoided and the number of variables is very large, making explicit modeling difficult, statistical matching has seldom been discussed in the context of multiple imputation. With the development of new computational methods and software, however, new techniques have become more attractive for researchers and practitioners. Thus, we discuss concepts of multiple imputation and their restrictions culminating in alternative approaches to solving the matching task and controlling its identification problems.

1.4 Identification Problems Inherent in Statistical Matching

Our objective is to learn which conclusions can and which cannot be drawn given specified combinations of assumptions and data. Addressing identification problems in the social sciences, Manski (1995) proposed estimating upper and lower bounds rather than producing sharp predictions for any parameter of interest. Concerning the problem of nonresponse, his idea is best illustrated by means of a simplified example. Consider a sample of n persons being asked a question with n_0 of them refusing to answer. Let the outcome variable U of interest have a value of either one or zero. Suppose that from p percent of the remaining $n_1 = n - n_0$ units a value of one for this variable U is observed. For estimating the probability $P(U = 1)$ of observing $U = 1$ often simply the percentage p is taken. This implies that the unobserved outcomes U have the same distribution as the observed ones. Manski (1995), pp. 21–31, on the other hand, argues that there is more to learn from the data. Consider p only as a "good" estimate of $P(U = 1|R = 1)$ whereas R is an indicator variable with a value of one if the outcome variable U is observed for a certain unit and zero if the answer is refused. In general

$$\begin{aligned} P(U=1) &= P(U=1, R=1) + P(U=1, R=0) \\ &= P(U=1|R=1)P(R=1) + P(U=1|R=0)P(R=0) \end{aligned}$$

holds but $P(U = 1|R = 0)$ is unknown. Using p as an estimate for $P(U = 1)$ assumes that the unknown quantity $P(U = 1|R = 0)$ is also best represented or estimated by p. However, we only know for sure $P(U = 1|R = 0) \in [0, 1]$; thus, we can derive estimates for the upper and lower bounds of $P(U = 1)$ with

$$\begin{aligned} P(U=1) &\geq P(U=1|R=1)P(R=1) \\ P(U=1) &\leq P(U=1|R=1)P(R=1) + P(R=0). \end{aligned}$$

Estimates of $P(R = 1)$ and $P(R = 0)$ are given by n_1/n and n_0/n, respectively. To achieve estimates of the bounds we weight p and calculate

$$p\frac{n_1}{n} \leq \widehat{P}(U=1) \leq p\frac{n_1}{n} + \frac{n_0}{n}.$$

Manski (1995), p. 27, takes pn_1/n as a lower bound estimate and $(pn_1 + n_0)/n$ as an upper bound (or worst, respectively, best case) estimate for the probability of interest.[2]

Notice that this sort of bounds is due to a "population concept" (see Manski (1995), p. 26), expressing what could be learned from those units that did not answer. The traditional confidence intervals on the contrary are based on sampling concepts and express the precision of an estimate obtained from a fixed sample. The discussion about the usefulness of these bounds in general is not without controversies; for a recent discussion see Horowitz and Manski (2000a and b) and Raghunathani (2000). However, we regard these bounds as especially useful in the context of statistical matching.

As we have already discussed, statistical matching initially is connected to an identification problem concerning the association of the variables not jointly observed. The conditional association (i.e., the association of the variables never jointly observed given the variables common to both files) cannot be estimated from the observed data; for a detailed proof see Rubin (1974). As already cited verbatim, Rubin (1987), p. 187, states that multiple imputation in the context of statistical matching may be used reasonably to display sensitivity for different choices of parameters of conditional association. We would like to go further. Transferring Manski's identification problems to the statistical matching task, we look at multiple imputation being the vehicle to achieve upper and lower bounds

[2] To be more general, estimating such bounds conditioning on various covariates is straightforward. We condition the estimation of such bounds on covariates hereinafter, possibly in a sense other than that intended by Manski (1995).

of the unconditional association of the variables never jointly observed, because we often are simply not able to calculate these bounds directly. See, for example, a recent frequentist approach of Moriarity and Scheuren (2001). Specifying admissible values of the inestimable association can be rather cumbersome. However, by means of multiple imputation based on explicit models, imputations can be made for different values describing the conditional association. From these imputed data sets the unconditional association can then be estimated and display its possible range. The idea behind that consideration is the use of suitable common variables determining the unconditional association as far as possible. In terms of regression analysis we would expect the common variables to be good predictors for the variables never jointly observed.

Consider, for example, a common variable Z determining another variable X only observed in one file. For demonstration purpose we discuss linear dependencies; i.e., let the correlation $\rho_{ZX} = 1$, and thus $X = a + bZ$ for some $a, b \in \mathbb{R}^2, b \neq 0$. The correlation between this common variable Z and a variable Y in a second file may be $\rho_{ZY} = 0.8$. It is easy to see that the unconditional correlation of the two variables X and Y which are not jointly observed is determined by Z with $\rho_{XY} = \rho_{a+bZ\,Y} = \rho_{ZY} = 0.8$. If the correlation between X and Z is less than one, say 0.9, we can easily calculate the possible range of the unconditional association between X and Y by means of the determinant of the covariance matrix which has to be positive (semi)definite; i.e., the determinant of the covariance matrix $cov(Z, X, Y)$ should be positive or at least zero. Given the above values and setting the variances to one without loss of generality, the covariance matrix of (Z, X, Y) is

$$\begin{aligned} cov(Z,X,Y) &= \begin{pmatrix} 1 & 0.9 & 0.8 \\ 0.9 & 1 & cov(X,Y) \\ 0.8 & cov(X,Y) & 1 \end{pmatrix} \quad \text{with} \\ \det(cov(Z,X,Y)) &= -cov(X,Y)^2 + 2 \cdot 0.72 cov(X,Y) - 0.45. \end{aligned}$$

Calculating the roots of $\det(cov(Z, X, Y))$ we get the two solutions $cov(X, Y) = 0.72 \pm \sqrt{0.0684}$. Hence we find the unconditional correlation bounded between [0.4585, 0.9815]; i.e., every value of the unknown covariance $cov(X, Y)$ greater than 0.4585 and less than 0.9815 leads to a valid and thus feasible covariance structure for (Z, X, Y). By means of the observed data we are not able to decide which covariance matrix could have generated the data. The $cov(X, Y)$ is nonidentifiable.

If the variables X and Y are conditionally independent or at least uncorrelated given Z, a correlation of X and Y of $cov(X, Y) = 0.72$ is computed which is exactly the middle of the interval [0.4585, 0.9815] yielding the maximum value for the determinant $|cov(Z, X, Y)|$. Finally we have found an upper (0.9815) and a lower (0.4585) bound for $cov(X, Y)$ according to Manski's population concept. Notice that estimating these bounds in the

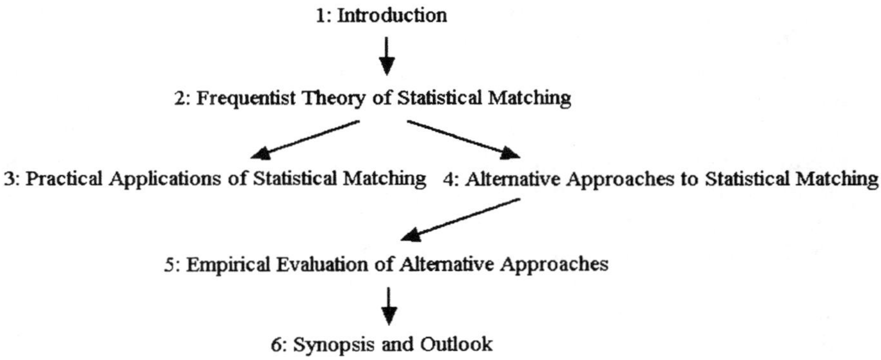

Figure 1.3. Skeleton of the book

matching context was first published by Kadane in 1978 (see the reprint Kadane (2001)). The strength of the conditional independence assumption was also discussed in a similar example by Rodgers (1984). He showed that only an extremely high correlation narrows the range of the unconditional association considerably.

Based on these bounds we propose a measure of the explanatory power of the common variables and, thus, of the validity of the statistical matching procedure. Since it is not possible to judge the quality of the matched data concerning this unobserved association whenever the variables are never jointly observed and no auxiliary data file is available, the range of these bounds may be used to achieve such a measurement. The less the bounds differ the better is the explanatory power of the common variables and the more valid results traditional matching techniques will produce. However, if the data structure is complex and high-dimensional we will not be able to calculate these bounds directly. It is shown later in this work that the multiple imputation principle allows us to fix the conditional association using prior information and provides a helpful and flexible tool to derive these bounds. We no longer have to stick to conditional independence when matching data from different sources is the task.

1.5 Outline of the Book

Basically the book is organized in six chapters. The skeleton of the book is described by Figure 1.3 showing the directed graph of its chapter headings. The graph indicates the order in which the chapters should be read.

Chapter 2: Frequentist Theory of Statistical Matching. This chapter points out the basic principles of statistical matching in detail. A frequentist theory is developed to analyze the properties of the matching process systematically. We explain how conditional independence is established by traditional statistical matching techniques. Objectives of statistical matching are derived, structured, and discussed in the light of their feasibility. Four levels of validity that a matching technique may achieve are introduced. Finally, the last section concerns conditional independence in general, and some simulation studies are presented for illustration purposes.

Chapter 3: Practical Applications of Statistical Matching. An overview of the roots and history of statistical matching and its applications is given herein. Statistical matching has a long tradition in European market research, especially in France and Germany where it is called data fusion, as well as in the USA and Canada. However, few papers have been published and available to the public until now, nor has an overview been available summarizing recent matching techniques. We try to bridge this gap and describe in detail some of the most popular matching techniques used in Germany and other countries for many years.[3]

Chapter 4: Alternative Approaches to Statistical Matching. This is the core chapter of the book. Here we introduce multiple imputation to the reader as proposed by Rubin (1987). Notational preliminaries concerning the treatment of missing data in general as well as statistical matching in particular are presented. Principles of Bayesian argumentation are explained on simplified matching examples first. Then imputation techniques such as data augmentation and iterative univariate imputations based on Markov chain Monte Carlo methods are discussed. A new noniterative multivariate imputation method especially suited for solving the matching task is derived successively. We discuss the use of prior information in detail to estimate bounds of the unconditional association. It is shown how the matching process may be improved by the use of a third complete data source. Finally, the alternative approaches proposed are compared by means of a simulation study, and their advantages and disadvantages are highlighted. We regard the results as rather encouraging. Statistical matching may be improved by these Bayesian-based techniques; moreover, they are quite helpful in overcoming the problem of conditional independence.

Chapter 5: Empirical Evaluation of Alternative Approaches. The practical importance of the alternative methods proposed is shown in this chapter. Real media data from market research are used to show extensively how these model-based approaches may be applied even if the data do not follow any simplifying modeling assumptions. Data problems that may typically occur in the majority of cases are discussed and solved as far as necessary. Three techniques are discussed in detail for matching the data. Prior information is used either assuming conditional independence

---

[3]Supported by the German market research company GfK in Nuremberg and Bureau Wendt in Hamburg and Puidoux, we got a closer look at many unpublished or scarcely documented procedures. Among others Statistics Canada was also very helpful with material.

or being estimated by another small sample of a suitably complete nature. Moreover, we derive some practical implications to recognize to what extent data may successfully be matched.

Chapter 6: Synopsis and Outlook. The book ends with a summary and conclusions of this work. A short overview of currently available software for statistical matching is given there also. Finally, an advance look at a path future work may follow is given.

Computer programs and routines as well as tables are listed in the appendix. The main routine developed in this book is described in the appendix in detail.

Each chapter starts with an introduction and a short outline of how the chapter is organized. At the end of each chapter all its results are summarized. Those readers only interested in an overview of the scope of a chapter or already familiar with the techniques discussed therein may read only the respective chapter's introduction and summary. Figure 1.3 shows the logical order in which the chapters are linked.

1.6 Bibliographic and Software Notes

A global overview of techniques for handling missing data discussing various approaches as well as the EM algorithm in detail is given by Little and Rubin (1987). A review of the current state of the field of nonresponse also describing weighting and imputation techniques is provided by Groves et al. (2002). Multiple imputation is introduced by Rubin (1987); some algorithms proposed there are extended herein. An excellent introduction to multiple imputation and data augmentation is given by Schafer (1997), a book which had great influence on this work. Further overviews of Markov chain Monte Carlo methods such as data augmentation and Gibbs sampling can be found in books by Tanner (1996), Gilks et al. (1996), Robert and Casella (1999), and Chen et al. (2000). General introductions to Bayesian inference are provided by Box and Tiao (1992), Gelman et al. (1995), and Carlin and Louis (2000).

A first matching methodology from the frequentist point of view appears in Goel and Ramalingam (1989). Koschnick (1995) contains a short description of a statistical matching technique as it has been typically applied in German market research over the years. A basic discussion about some statistical properties of the statistical matching process concerning conditional independence was published by Rodgers (1984). There we find the results of other simulation studies presented and summarized. Conditional independence is particularly addressed by Whittaker (1990) as well as Cox and Wermuth (1996).

Some algorithms described in this book have been implemented by the author for general use as functions in the statistical language S-PLUS

4.5, copyrighted by MathSoft, Inc. Information on the routines is provided in the appendix. All computations are basically performed with S-PLUS 4.5 partly using special libraries which are also available for free. The standalone version NORM provided by Schafer (1999b) is also used for some computations. An overview of software handling missing data problems may be found by the interested reader via the Internet Web site www.multiple-imputation.com.

2

Frequentist Theory of Statistical Matching

To the uninitiated data fusion might sound like major market research fraud. How do you react to the proposition that two different sample surveys covering different subject areas with different respondents can be statistically joined together? The respondents in the resulting data set will have all the answers to all the questions in both the original surveys. Your first reaction will probably be cynical disbelief.

K. Baker, P. Harris, J. O'Brien (1989)

2.1 Introduction and Chapters Outline

The objective of statistical matching techniques is the generation of a new data set that allows even more flexible analysis than each single data set. In particular, the associations between variables never jointly observed are specified in such a completed data set. In this chapter we show whether a statistically matched file may be analyzed as if it were a single sample.

Especially in the area of media analyses, statistical matching has been performed in France, the United Kingdom, and in Germany with a reasonable degree of validity as published by Antoine (1987), Antoine and Santini (1987), Scheler and Wiegand (1987), Baker et al. (1989), Baker (1990), Roberts (1994), Adamek (1994), and Wendt (1976, 1979, 1980, 1986 and 2000). Further descriptions of statistical matches done in the United States were written by Okner (1972a and b, 1974), Ruggles and Ruggles (1974), and Ruggles et al. (1977); for an overview, see Rubin (1986). Alter (1974) and later Liu and Kovacevic (1997, 1998) explained the merging of data sets performed in Canada.

However, there is skepticism among statisticians about the power of matching techniques; most famous are the comments of Sims (1972a and b) and the evaluation and overview given by Rodgers (1984). The assumption of conditional independence is often stressed in the literature of statistical matching. Ruggles et al. (1977), Woodbury (1983), Rodgers (1984), Bennike (1987), Goel and Ramalingam (1989), Kovacevic and Liu (1994), and Gabler (1997) are known to mathematically study matching algorithms and investigate their validity under certain circumstances. For example, they give some empirical evidence that violation of the conditional independence assumption may result in large estimation errors. Paass (1986) suggests the use of some additional information from another data source. Using auxiliary information as an alternative to the conditional independence assumption is also studied by Singh et al. (1988, 1993). They introduce additional constraints in terms of auxiliary categorical tables to improve the quality of the matched file. Liu and Kovacevic (1996, 1998) extend this approach and perform an extensive simulation study based on some real data provided by Statistics Canada.

In this chapter we discuss and derive a frequentist framework for statistical matching. This enables us to explain why conditional independence of the variables never jointly observed given the common variables will be established by any matching technique that uses nearest neighbor matches. Moreover, we show that even matching based on the propensity scores will establish conditional independence but with a loss in validity concerning the preservation of the joint distributions. Based on the conditional independence assumption and the scale of the variables available, different levels of validity of statistical matching are formulated and structured. The extent to which the power of a statistical matching technique may be measured is also discussed. A "minimum" validity is then postulated. Finally, some applications of the derived framework to specific distributions are shown and a simulation study is presented for the purpose of illustration.

2.2 The Matching Process

Statistical matching is initiated by two (or more) samples, one usually of larger size than the other, with the number of individuals appearing in both samples (i.e., the overlap) clearly negligible. Only certain variables (denoted by Z) of the interesting individual's characteristics can be observed in both samples; they are called common variables. Some variables Y appear only in one sample whereas other variables X are observed exclusively in the second sample; X and Y are called specific variables. For the purpose of generalization X, Y, and Z can be regarded as vectors of variables. In media

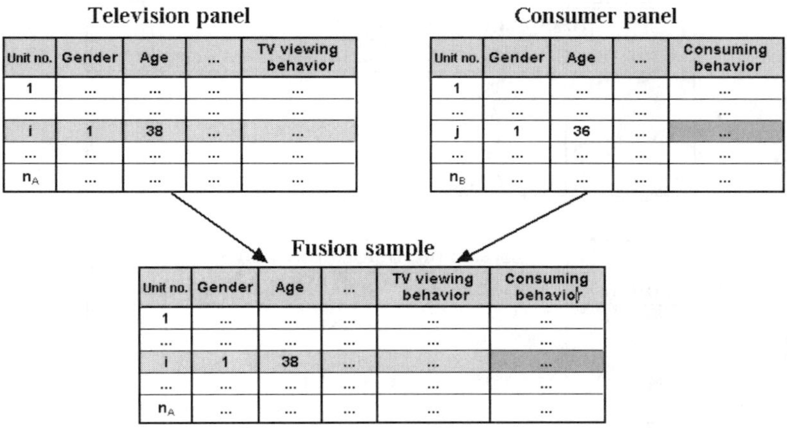

Figure 2.1. Principle of statistical matching

practice social class, housing conditions, marital status, terminal age of education, education, and many other variables as well as gender and age would be used as Z variables for a linking mechanism. Since usually no single sample exists with information on X, Y, and Z together, an artificial sample has to be generated by matching the observations of both samples according to Z; see Figure 2.1 for illustration.

The objective of statistical matching is the creation of a complete "microdata" file where every unit provides observations of all X, Y, and Z variables. Therefore the matching is performed at an individual level by means of statistical matching; this is often called the marriage process. The individual level may apply to households, to each individual living in a household, or to establishments. Notice that the variables observed in both samples should be measured at the same level; at least, they should be aggregated to the same level. For instance, the weekly purchases of a household should be added to the media consumption of another household, and not to the TV viewing information of an individual.

Without loss of generality, let the (X, Z) sample be the recipient sample A of size n_A and the (Y, Z) sample the donor sample B of size n_B. For every unit i, $i = 1, 2, \ldots, n_A$, of the recipient sample with the observations (x_i, z_i) a value y from the observations of the donor sample is determined and a data set $(x_1, y_1, z_1), \ldots, (x_{n_A}, y_{n_A}, z_{n_A})$ with n_A elements of the recipient sample is constructed. The main idea is to search for a statistical match, i.e., for a donor unit j with $(y_j, z_j) \in \{(y_1, z_1), (y_2, z_2), \ldots, (y_{n_B}, z_{n_B})\}$ whose observed data values of the common variables z_j are identical to those z_i of the recipient unit i for $i = 1, 2, \ldots, n_A$. Figure 2.2 displays the notation we use to describe the statistical matching procedure.

Notice that $\widetilde{y}_i$ is not the true y-value of the ith recipient unit but the y-value of the matched statistical twin. In the following all density func-

Recipient sample before matching			
Unit no.	Variable X	Variable Z	Variable Y
1	x_1	z_1	
2	x_2	z_2	
...	...	...	
n_A	x_{n_A}	z_{n_A}	

Donor sample before matching			
Unit no.	Variable X	Variable Z	Variable Y
1		z_1	y_1
2		z_2	y_2
...		...	...
n_B		z_{n_B}	y_{n_B}

Recipient sample after matching			
Unit no.	Variable X	Variable Z	Variable Y
1	x_1	z_1	$\widetilde{y}_1$
2	x_2	z_2	$\widetilde{y}_2$
...	...	...	...
n_A	x_{n_A}	z_{n_A}	$\widetilde{y}_{n_A}$

Figure 2.2. Samples to be operated by statistical matching

tions (joint, marginal, or conditional) and their parameters produced by the matching algorithm are marked by the symbol $\tilde{}$.

The question of which sample should be used as the recipient sample cannot be answered in general and depends on the particular matching task. Often the larger sample is taken as the recipient sample because no information should be discarded. In this case the multiple use of donor units is necessary.

We have assumed that the overlap is poor and matching of identical units is impossible. To find a perfect match for each individual in terms of the common variables Z may also be impossible, especially if (some) common variables are continuous. Therefore, as described, e.g., by Baker (1990), Roberts (1994), and Antoine (1987), the marriage process is often carried out using an algorithm based on nearest neighbor techniques calculated by means of a distance measure $d(.,.)$. The marriage algorithm may use all or some of the common variables, weighted or not, to find for each recipient unit i, $i = 1, 2, \ldots, n_A$, at least one donor unit j from $j = 1, 2, \ldots, n_B$, whose distance $d(z_i, z_j)$ is minimal.

For example, in German market research it is common practice to use distance functions $d(.,.)$ based either on a quadratic distance (Euclidean) or an absolute distance (City-Block); see Arbeitsgemeinschaft Media-Analyse e.V. (1996). To account for the different scales of the common variables it is recommended to standardize continuous and even ordinal variables to a mean of zero and a standard error of one. Categorical variables are recoded into zero and one for each category. Also an individual weight $g(Z)$ could be given for each common variable Z to account for its importance for the matching process. In a recent matching project some sociodemographic variables are used as common variables Z, such as Z_1 : gender (0/1, weight 15), Z_2 : age (standardized, weight 10), $Z_3 - Z_7$: working status (5 categories, recoded into 0/1, weight 3 each), $Z_8 - Z_{12}$: education (5 categories, recoded into 0/1, weight 3 each), Z_{13} : dogs in the household (0/1, weight 2.5), Z_{14} : computer in the household (0/1, weight 2), and so on. Based on these, say p variables $Z_1, Z_2, \ldots, Z_p$, the weighted distance given by

$$d(z_i, z_j) = \sqrt{\sum_{k=1}^{p} g(z_k)(z_{ki} - z_{kj})^2}\,, \tag{2.1}$$

$i = 1, 2, \ldots, n_A$, $j = 1, 2, \ldots, n_B$ for every pair of the donor and the recipient sample may be calculated; for more details see Section 3.3.5. Finally, for every unit i, $i = 1, 2, \ldots, n_A$, of the recipient sample all the observed information of variables Y of the donor unit j, $j = 1, 2, \ldots, n_B$, which has the smallest distance to i is imputed at once. We call such a procedure a nearest neighbor match. Often some of the common variables are treated as critical variables; i.e., the match between donor and recipient unit concerning these critical variables has to be perfect. Typically variables such as gender or region are regarded as critical variables. Such rules of exact matching are identical to a separation of the data and a statistical match within each group. A recent reference to a variety of nearest neighbor and hot deck procedures in the context of large-scale imputation for complex surveys is, e.g., Marker et al. (2002).

Donor units may be used for different recipient units arbitrarily often; such a marriage algorithm is called polygamy. By restricting the multiple choice of a donor for different recipients (e.g., bigamy or monogamy), further variations of the algorithm can be created. To limit the number of times a donor is taken, a penalty weight can also be placed on donors already used; e.g., any arbitrary value may be added to all distance functions (2.1) that contain this donor. The multiple choice of donor units will otherwise reduce the effective sample size and lead to underestimation of the true variance. Another modification is to take the next three (or any other number) nearest neighbor units and impute their (weighted) mean. This method also leads to a loss in variability. If the multiple use of donors is restricted or combined with a penalty function, the resulting artificial sample (i.e., the statistically matched file) may vary depending on the order in which the donor units are taken. To limit this problem some algorithms, for instance, allow abandoning certain matches in order to find a better donor–recipient combination afterwards. For a short description of such optimization algorithms see Antoine (1987).

There are many different matching programs and the most important difference is the matching algorithm itself. Many institutes have well-developed rules to decide which is the best match for each recipient. It often is a highly complex statistical and computational issue and even more difficult the higher the number of common variables. A more detailed description of different traditional approaches is given in Chapter 3. In the following sections we derive a mathematical framework for statistical matching. This enables us to discuss the results produced by statistical matching based on exact matches concerning the common variables as well as matches of nearest neighbor units.

2.3 Properties of the Matching Process

2.3.1 Basic Assumptions

Let X, Y, and Z be multivariate random variables with a joint probability or density function $f_{X,Y,Z}$. For discrete variables, $f_{X,Y,Z}(x_i, y_i, z_i)$ describes the probability of drawing a certain unit i with observation (x_i, y_i, z_i). For continuous variables $f_{X,Y,Z}(x_i, y_i, z_i)$ is the value of the joint density function at the point (x_i, y_i, z_i). Notice that $f_{X,Y,Z}$ may refer even to any plausible mixture of density functions. Without loss of generality we often simply use the expression "probability" hereinafter. In case of continuous variables, instead of the probability function f may be used as the density function.

We assume that the units of each sample as well as the samples themselves are drawn independently of each other and the same sampling design is applied to select the samples. Of course, this is a simplifying assumption but usually the samples to be matched are regarded as being representative of the population of interest; e.g., see O'Brien (1991). Much effort is put into correcting and weighting the data to achieve representative samples. According to the matching rules defined by Arbeitsgemeinschaft Media-Analyse e.V. (1996), for instance, ensuring that the two separate samples do not differ significantly by means of the common variables is recommended. Thus, our assumption may reflect what is already assumed by practitioners.

Now the distribution of the donor sample of size n_B is given by $\prod_{i=1}^{n_B} f_{Y,Z}(y_i, z_i)$ as is the recipient sample of size n_A distributed according to $\prod_{i=1}^{n_A} f_{X,Z}(x_i, z_i)$. Furthermore, let the matching algorithm be one of multiple choice of the donor units without any penalty function. Hence the units of the artificial sample with observations $(x, \widetilde{y}, z)$ can also be regarded as being drawn independently of each other, each with distribution $\widetilde{f}_{X,Y,Z}(x, y, z)$. The matching algorithm induces the distribution

$\prod_{i=1}^{n_A} \widetilde{f}_{X,Y,Z}(x_i, y_i, z_i)$ on the set of all possible statistically matched files. They can be handled as simple random samples drawn from an artificial population with distribution $\widetilde{f}_{X,Y,Z}(x, y, z)$, which we call the artificial distribution.

Often the statistically matched file is used to estimate parameters (such as means, variances, covariances, or higher moments) of the "initial" population following $f_{X,Y,Z}(x, y, z)$ with traditional methods. To judge the quality of such estimates, which in fact means the power of the match, the relation between $\widetilde{f}_{X,Y,Z}(x, y, z)$ and $f_{X,Y,Z}(x, y, z)$ has to be examined.

2.3.2 Distribution of the Artificial Sample

Without loss of generality, let the probability of drawing a particular unit i, $i = 1, 2, \ldots, n_A$, after the matching with observation (x_i, y_i, z_i) be described by $\widetilde{f}_{X,Y,Z}(x_i, y_i, z_i)$. This is equivalent to the probability of drawing a particular unit i of the recipient sample with observation (x_i, z_i) and merging this unit with a unit j from the donor sample with observed values (y_j, z_j), where $z_i = z_j$. The probability of a donor unit j with observed value $z_j = z_i$ from Z to have the observation y_j from Y is obviously $f_{Y|Z}(y_j|z_j)$. Hence the probability of observing (x, y, z) for any unit of the statistically matched file is

$$\widetilde{f}_{X,Y,Z}(x, y, z) = f_{X,Z}(x, z) f_{Y|Z}(y|z)\,,$$

provided that the units of the donor and recipient sample have been drawn independently from the same population. The same argument holds if the variables are continuous and $\widetilde{f}_{X,Y,Z}(x, y, z)$ and $f_{X,Y,Z}(x, y, z)$ refer to a density instead of a probability function. Remember that the direction of the matching is chosen arbitrarily; we may impute X to the (Y, Z) sample as well. Clearly, the empirical distributions derived from the recipient sample are never altered by any matching technique. However, we want to abstract from the available observed samples and derive conditions under which any statistically matched file may be regarded as a random sample drawn from the actual population of interest.

In general, the artificial distribution produced by the statistical match is

$$\begin{aligned} \widetilde{f}_{X,Y,Z}(x, y, z) &= f_{X,Z}(x, z) f_{Y|Z}(y|z) \\ &= f_{X|Z}(x|z) f_Z(z) f_{Y|Z}(y|z) \\ &= f_{X|Z}(x|z) f_{Y,Z}(y, z) \end{aligned} \tag{2.2}$$

and the conditional matching distribution is given by

$$\widetilde{f}_{X,Y|Z}(x, y|z) = f_{X|Z}(x|z)\; f_{Y|Z}(y|z)\,. \tag{2.3}$$

It should be noted that this derivation is only admissible if, for every recipient unit, there is a donor unit with the same observed value z for Z. Especially when the Z variables are continuous or the number of available

Z variables is large, an exact match will seldom be found and a nearest neighbor unit in z has to be merged. The influence of such a nearest neighbor match on the artificial distribution in the statistically matched file (i.e., the fusion distribution) as derived in (2.2) is hereinafter discussed on the basis of simulation studies.

2.3.3 Marginal Distributions After the Match

The marginal distributions of $\tilde{f}_{X,Y,Z}$ are now easily obtained with the help of (2.2):

$$\begin{aligned}\tilde{f}_X(x) &= \int\int \tilde{f}_{X,Y,Z}(x,y,z)dydz \\ &= \int f_{X,Z}(x,z)\int f_{Y|Z}(y|z)dydz = f_X(x) \qquad (2.4)\end{aligned}$$

as is $\tilde{f}_Y(y) = f_Y(y)$ and $\tilde{f}_Z(z) = f_Z(z)$. Furthermore

$$\begin{aligned}\tilde{f}_{X,Z}(x,z) &= \int \tilde{f}_{X,Y,Z}(x,y,z)dy \\ &= f_{X,Z}(x,z)\int f_{Y|Z}(y|z)dy = f_{X,Z}(x,z) \qquad (2.5)\end{aligned}$$

and also $\tilde{f}_{Y,Z}(y,z) = f_{Y,Z}(y,z)$. The following are different from their initial distributions.

$$\begin{aligned}\tilde{f}_{X,Y}(x,y) &= \int \tilde{f}_{X,Y,Z}(x,y,z)dz = \int f_{X|Z}(x|z)f_Z(z)f_{Y|Z}(y|z)dz, \\ \tilde{f}_{X,Y,Z}(x,y,z) &= f_{X,Z}(x,z)f_{Y|Z}(y|z) \\ &= f_{X,Z}(x,z)f_{Y|X,Z}(y|x,z)\,\frac{f_{Y|Z}(y|z)}{f_{Y|X,Z}(y|x,z)} \\ &= f_{X,Y,Z}(x,y,z)\,\frac{f_{Y|Z}(y|z)}{f_{Y|X,Z}(y|x,z)} \\ &= f_{X,Y,Z}(x,y,z)\,\frac{f_{X|Z}(x|z)}{f_{X|Y,Z}(x|y,z)}. \qquad (2.6)\end{aligned}$$

Thus, the distribution of X, Y, and Z after the match is equal to the initial distribution, if X and Y are independent conditional on every possible value z of Z; i.e.,

$$f_{Y|X,Z}(y|x,z) = f_{Y|Z}(y|z) \quad \text{or equivalent} \quad f_{X|Y,Z}(x|y,z) = f_{X|Z}(x|z).$$

Sims (1972a and b) was the first to point out conditional independence as a useful assumption for statistical matching. By means of this frequentist framework we see that conditional independence is the main requirement for a reliable match when inference is intended about the joint distribution of the variables never jointly observed. In practice we often find this assumption of conditional independence formulated, e.g., as a "strong predictive relationship between common matching variables and recipient-donor measures" (Adamek (1994)).

Moreover, all marginal distributions, which could have been estimated already by the two separate samples, are identical before and after matching. Only the joint distributions of variables not jointly observed may be different. Note that all derivations above are valid for vectors of random variables X, Y, and Z as well. All moments concerning variables of one or the other sample are identical for the artificial distribution in the statistically matched file and the initial distribution. In general, we get

$$\begin{aligned} \widetilde{E}(g(X)) &= E(g(X)), \quad \widetilde{E}(g(Y)) = E(g(Y)), \quad \widetilde{E}(g(Z)) = E(g(Z)), \\ \widetilde{E}(g(X,Z)) &= E(g(X,Z)), \text{ and } \widetilde{E}(g(Y,Z)) = E(g(Y,Z)) \end{aligned} \tag{2.7}$$

for any measurable function g, existence of moments assumed. We see by (2.7) that a test of the power of the match based on the marginal distributions of the variables is not very helpful although so far it is the only possibility for controlling the validity of a matching technique. In practice the empirical distributions of the common variables Z as well as the imputed variables Y in the resulting statistically matched file are often compared with their empirical distributions of the donor sample to evaluate the "similarity" of both samples. Usually a statistical match is said to be successful if the marginal and joint empirical distributions of Z and Y as they are observed in the donor sample are nearly the same in the statistically matched file; see Wendt (1976, 1979), Baker et al. (1989), and Koschnick (1995). These aspects are discussed in Section 2.5.

2.3.4 Correlations After the Match

To proceed further we consider the moments of the joint distribution of variables referring to different samples. The association between X and Y generated by the match, measured as covariance $\widetilde{cov}(X,Y)$, is given by

$$\widetilde{cov}(X,Y) = cov(X,Y) - E(cov(X,Y|Z)). \tag{2.8}$$

This result can easily be obtained by calculating $E(cov(X,Y|Z))$:

$$\begin{aligned} E(cov(X,Y|Z)) &= \int [E(XY|Z=z) - E(X|Z=z)E(Y|Z=z)]\, f_Z(z)dz \\ &= \int \Big[\int\int xy f_{X,Y|Z}(x,y|z)dxdy \\ &\quad - \int x f_{X|Z}(x|z)dx \int y f_{Y|Z}(y|z)dy \Big] f_Z(z)dz \\ &= \int\int\int xy f_{X,Y|Z}(x,y|z) f_Z(z)dxdydz \end{aligned}$$

$$\begin{aligned} & -\int\int\int xyf_{X|Z}(x|z)f_{Y|Z}(y|z)f_Z(z)dxdydz \\ = & \int\int\int xyf_{X,Y,Z}(x,y,z)dxdydz \\ & -\int\int\int xy\widetilde{f}_{X,Y,Z}(x,y,z)dxdydz \\ = & E(XY) - \widetilde{E}(XY) \\ = & E(XY) - \mu_X\mu_Y - (\widetilde{E}(XY) - \mu_X\mu_Y) \\ = & cov(X,Y) - \widetilde{cov}(X,Y) \text{ since } \mu = \widetilde{\mu}. \end{aligned}$$

Because generally

$$cov(X,Y) = E(cov(X,Y|Z)) + cov(E(X|Z), E(Y|Z)) \qquad (2.9)$$

holds (see Whittaker (1990)) p. 125, we get from (2.8) and (2.9)

$$\widetilde{cov}(X,Y) = cov(E(X|Z), E(Y|Z)). \qquad (2.10)$$

The best prediction of X or Y in the linear model conditioned on $Z = z$ is known to be its conditional expectation $E(X|Z = z)$ or $E(Y|Z = z)$, respectively; see Mood et al. (1974), p. 491. In this case, the covariance of X and Y in the matched data set is equal to the covariance of their best linear predictions based on the common variables Z of the single data sets.

Notice that the expected or average covariance of X and Y given $Z = z$ is just the difference of the covariances from the initial and the fusion distributions, i.e.,

$$E(cov(X,Y|Z)) = cov(X,Y) - \widetilde{cov}(X,Y). \qquad (2.11)$$

When there is access to a small but completely observed third data source then $E(cov(X,Y|Z))$ may be estimated therefrom and used as a quality measurement of the match. The closer this value gets to zero, the better the true correlation is reproduced by the fusion.

Hence, statistical matching can produce good results concerning the true correlation between the variables X and Y never jointly observed only if they are on the average conditionally uncorrelated; i.e., $E(cov(X,Y|Z)) = 0$. The independence of X and Y conditional on $Z = z$ is sufficient but not necessary therefore. If the interest is not in displaying the true distribution $f_{X,Y,Z}$ in the statistically matched file but the estimation of moments therefrom, then X and Y should be, at least, on the average conditionally uncorrelated to achieve the usual frequency valid estimates of the moments.

2.4 Matching by Propensity Scores

Propensity scores are often used in observational studies to generate suitable control groups which are in some way similar to treated groups when a randomized experiment is not available; see Rosenbaum and Rubin (1983, 1985), Holland (1986), and Rubin and Thomas (2000). Propensity score matching refers to a number of multivariate methods to construct treated and control samples that have similar distributions on many variables; these are the so-called covariates. This is of special interest when the treatment group is small and quite different in its covariates compared to a larger group not being treated. In this context the propensity scores are helpful for extracting suitable units from the larger group to build up a control group allowing us to more or less map an experimental situation. Propensity score matching is regarded as a statistical solution to the fundamental problem of causal inference, since it is not possible to measure the effect of treatment and no treatment on one single unit; see Holland (1986) or Rosenbaum (1999). Roughly speaking, the difference of the means of the interesting treatment variable based on treated units and control units created as statistical twins is analyzed via propensity score matching.

Contrary to this, in the context of statistical matching our objective is to show whether and to what extent propensity score matching may be used instead of a nearest neighbor match. We show that the exact match described before is a special case of propensity score matching.

2.4.1 Principle

First a simple example illustrates the procedure of propensity score matching. We extend the recipient sample (X, Z) by an outcome variable S, with $S_i = 1$ for all units $i = 1, 2, \ldots, n_A$ of this sample. The variable S is also added to the donor sample with $S_i = 0$ for $i = 1, 2, \ldots, n_B$; for illustration see Figure 2.3.

Putting both samples together, a logit or a probit model is estimated with one dependent variable S and the common variables Z as independent variables, including the constant of a regression. Neither X nor Y is used

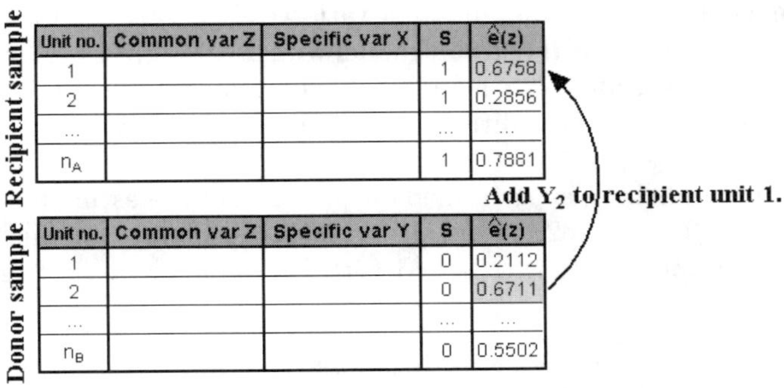

Figure 2.3. Principle of propensity score matching

throughout this estimation and matching procedure. The propensity score

$$e(z_i) = P(S = 1|Z = z_i) = g(z_i'\beta)$$

is defined as the conditional probability of a unit i, $i = 1, 2, \ldots, n$, $n = n_A + n_B$, to belong to a certain (treatment) group given the covariates $Z = z$. Based on the well-known formulas (see Greene (1997), p. 874),

$$\begin{aligned} \widehat{e}(z_i) = g(z_i'\widehat{\beta}) &= \frac{1}{1 + e^{-z_i'\widehat{\beta}}}, \quad \text{(logit) or} \\ &= \Phi(z_i'\widehat{\beta}) \quad \text{(probit)} \quad i = 1, 2, \ldots, n \end{aligned}$$

(Φ denotes the cumulative density function of the standard normal distribution), the individual propensity scores $e(z)$ are estimated from the regression output for β. To derive the maximum likelihood (ML) estimates $\widehat{\beta}$ of the regression coefficient β a Bernoulli distribution for S_i is assumed with parameter $p_i = P(S = 1|Z = z_i) = e(z_i)$; that is,

$$S_i \sim \text{Bernoulli}(p_i), \quad i = 1, 2, \ldots, n.$$

Then the joint distribution of n independent observations from S is given by

$$f_{S_1,\ldots,S_n|Z_1,\ldots,Z_n}(s_1, \ldots, s_n|z_1, \ldots, z_n) = \prod_{i=1}^{n} e(z_i)^{s_i}(1 - e(z_i))^{1-s_i} \quad (2.12)$$

which defines the likelihood of the model. Let us assume the existence of a latent variable $S^\star$ following a logistic distribution with mean $z_i'\beta$ and variance $\pi^2/3$. If $S_i = 1$ is observed only when $S_i^\star > 0$, $i = 1, 2, \ldots, n$, then the probability for $S = 1$ is given by

$$\begin{aligned} P(S_i = 1|Z = z_i) &= P(S_i^\star > 0|Z = z_i) \\ &= \frac{1}{1 + e^{-z_i'\beta}} = e(z_i), \quad i = 1, 2, \ldots, n. \quad (2.13) \end{aligned}$$

Assuming a normally distributed latent variable $S_i^\star \sim N(z_i'\beta, 1)$ with $S_i = 1$ only when $S_i^\star > 0$, yields the probit model with $P(S = 1|Z = z_i) = \Phi(z_i'\beta)$, $i = 1, 2, \ldots, n$. Within such a logit or probit model the ML estimates often cannot be calculated directly, therefore an iterative algorithm like the Newton–Raphson may be used.[1]

Finally, the matching is performed on the basis of the estimated propensity scores $\widehat{e}(z_i)$, $i = 1, 2, \ldots, n_A + n_B$. For every recipient unit i, $i = 1, 2, \ldots, n_A$, a donor unit is searched with the same or nearest estimated

[1]For further details of the logistic regression or the probit regression model and their estimation procedures see Greene (1997), pp. 873–894. The necessary calculations can easily be performed using standard econometric software, for example, by means of the software package LIMDEP 7.0 provided by Econometric Software, Inc.

propensity score. If such a donor unit is found for every recipient unit, then its y-values are added to the recipient unit.

2.4.2 Balancing Scores

We have already seen in the former sections that exact matches in the common variables Z will yield to conditional independence. If the number of common variables is large, the calculation of the exact or nearest neighbor matches may be cumbersome, especially when different weights are applied to the common variables. Moreover, the separate samples may show different empirical distributions of the common variables due to, for example, oversampling special population groups or different sampling designs. Especially in such cases propensity score matching is proposed as an attractive alternative.

A generalized form, the balancing score, is discussed first. Let S again be the indicator variable with $S_i = 1$ if any unit i, $i = 1, 2, \ldots, n$, belongs to the (treated) recipient sample and $S_i = 0$ if unit i belongs to the (control) donor sample. In general, a balancing score $b(z)$ is given by a function b of the observed covariates Z such that the conditional distribution of Z given $b(z)$ is the same for recipient and donor units. Then Z and S are conditionally independent given the score with $f_{Z|S,b(z)} = f_{Z|b(z)}$; see Rosenbaum and Rubin (1983). Given identical values of the balancing scores in both samples, the distribution of the covariates is identical. In terms of probability distributions the probability of drawing a unit i, $i = 1, 2, \ldots, n_A$, with observation z_i in the recipient sample is the same as the probability of drawing a unit j from $j = 1, 2, \ldots, n_B$ with observation z_j in the donor sample, if $b(z_i) = b(z_j)$ holds. According to Rosenbaum and Rubin (1983) there are many functions that are balancing scores, and even estimates of balancing scores behave as do balancing scores. The most trivial and "finest" balancing score is Z itself; i.e., $b(z) = z$. Then the balancing score matching is identical to an exact match based on identical common variables as we have assumed in Section 2.3.2. Moreover, Rosenbaum and Rubin (1983) prove that the most general and "coarsest" function that is still a balancing score is the propensity score; i.e., $b(z) = e(z)$.

2.4.3 Propensity Scores

We continue to assume that the samples are drawn randomly and independently from the underlying population. Every unit in the population has a (theoretical) chance of being sampled. It is a randomized experiment and in terms of Rosenbaum and Rubin (1983) the assignment of the units to each sample is strongly ignorable given a vector of covariates $Z = z$. An assignment is strongly ignorable given the covariates $Z = z$ if the variables observed only in one sample are conditionally independent from the assignment to this sample given the covariates $Z = z$. Since we have assumed that the separate samples are random samples from the underlying population, $f_{X|S,Z} = f_{X|Z}$ and $f_{Y|S,Z} = f_{Y|Z}$ holds in our matching case. According to Rosenbaum and Rubin (1983), if the assignment is strongly ignorable given $Z = z$ then it is also strongly ignorable given any balancing score $b(z)$. Thus, for statistical matching $f_{X|S,e(z)} = f_{X|e(z)}$ and $f_{Y|S,e(z)} = f_{Y|e(z)}$ is also valid for the propensity scores $e(z)$.

To derive the joint distribution of the propensity score $e(Z)$ and the variables X and Y, let $e(Z)$ be considered the transformation of the random vector Z calculated according to (2.13) with $\beta \neq 0$. Then

$$\begin{aligned} u_1 &= g_1(z_1, z_2, \ldots, z_k) = e(z), \\ u_2 &= g_2(z_1, z_2, \ldots, z_k) = z_2, \\ &\ldots \\ u_k &= g_k(z_1, z_2, \ldots, z_k) = z_k \end{aligned}$$

is a one-to-one transformation. The inverse function $g^{-1} = (g_1^{-1} \ \ldots \ g_k^{-1})'$ of $g = (g_1 \ \ldots \ g_k)'$ is given by

$$\begin{aligned} z_1 &= g_1^{-1}(u_1, u_2, \ldots, u_k), \\ &\ldots \\ z_k &= g_k^{-1}(u_1, u_2, \ldots, u_k). \end{aligned}$$

According to the rules of variable transformation (see Casella and Berger (1990), pp. 176–177), the distribution of the propensity score $e(Z)$ is given by

$$\begin{aligned} & f_{e(Z)}(u_1) = f_{U_1}(u_1) \\ =& \int \cdots \int f_{U_1,\ldots,U_k}(u_1, \ldots, u_k) du_2 \cdots du_k \qquad (2.14) \\ =& \int \cdots \int f_{Z_1,\ldots,Z_k}(g_1^{-1}(u_1, \ldots, u_k), \ldots, g_k^{-1}(u_1, \ldots, u_k))|J| du_2 \cdots du_k, \end{aligned}$$

where $|J|$ denotes the determinant of the Jacobian matrix of the transformation with

$$J = \begin{pmatrix} \frac{\partial g_1^{-1}}{\partial u_1} & \cdots & \frac{\partial g_1^{-1}}{\partial u_k} \\ \cdots & \cdots & \cdots \\ \frac{\partial g_k^{-1}}{\partial u_1} & \cdots & \frac{\partial g_k^{-1}}{\partial u_k} \end{pmatrix}.$$

By means of variable transformation all the distributions $f_{Y,e(Z)}$, $f_{X,e(Z)}$, and $f_{X,Y,e(Z)}$ can be calculated theoretically, but the derivation of these distributions will often be cumbersome. However, (2.14) already indicates that the distribution $f_{e(Z)}$ of the propensity score $e(Z)$ may be quite different from the actual distribution f_Z of Z and we may not be able to recalculate f_Z therefrom. Analogous to our argument presented in Section 2.3.2 we achieve the fusion distribution $\tilde{f}_{X,Y,e(Z)}(x, y, e(z))$ by the product of the probability $f_{X,e(Z)}(x, e(z))$ of drawing a unit i with observations $(x_i, e(z_i))$

of the recipient sample and the probability $f_{Y|e(Z)}(y|e(z))$ of drawing a unit j with observations $(y_j, e(z_j))$ of the donor sample where the propensity scores are $e(z_i) = e(z_j)$. Since the samples are drawn independently of each other, the artificial distribution produced by statistical matching is given by

$$\begin{aligned}\tilde{f}_{X,Y,e(Z)}(x,y,e(z)) &= f_{X,e(Z)}(x,e(z))f_{Y|e(Z)}(y|e(z)) \\ &= f_{X|e(Z)}(x|e(z))f_{Y,e(Z)}(y,e(z)), \qquad (2.15)\end{aligned}$$

and the conditional fusion distribution for any unit of the statistically matched file is

$$\tilde{f}_{X,Y|e(Z)}(x,y|e(z)) = f_{X|e(Z)}(x|e(z))f_{Y|e(Z)}(y|e(z)).$$

Again the fusion distribution $\tilde{f}_{X,Y,e(Z)}$ produced by propensity score matching is only identical to the original distribution $f_{X,Y,e(Z)}$ if X and Y are conditionally independent given the propensity scores $e(Z) = e(z)$. Since X and Y are not transformed, their marginal distributions are still the same after statistical matching; i.e.,

$$\begin{aligned}\tilde{f}_X(x) &= \int\int \tilde{f}_{X,Y,e(Z)}(x,y,e(z))dy\ de(z) \\ &= \left(\int f_{X,e(Z)}(x,e(z)) \int f_{Y|e(Z)}(y|e(z))\right) dy\ de(z) \\ &= f_X(x),\end{aligned}$$

and $\tilde{f}_Y(y) = f_Y(y)$. By means of (2.15) we see that the joint distributions of X, Z and Y, Z are also affected by the matching procedure, as long as the finest balancing score $b(Z) = Z$ is not used. Thus we conclude that the coarser the balancing score the less accurate is the matching procedure.

We have assumed that the matching is performed via identical propensity scores. Usually not exactly the same propensity values are computed and again nearest neighbor values have to be taken; see Figure 2.3. Also we may restrict the multiple use of donor units. Many variations may be realized and further investigations about their validity are necessary.

2.5 Objectives of Statistical Matching

We have already discussed that the general benefit of statistical matching is the creation of one complete data source containing information about all variables. Without focusing on a special matching algorithm we suggest distinguishing the four levels of validity a matching procedure may achieve. We use the term validity rather than efficiency, because efficiency usually refers to a minimum mean squared error criterion as it is common, for example, in survey sampling theory and not to different levels of reproduction and preservation of the original associations and distributions.

- **First level:**
  The true but unknown values of the (multivariate) Y variable of the recipient units are reproduced; i.e., $\widetilde{y}_i = y_i$ for $i = 1, 2, \ldots, n_A$. We call this situation a "hit" for any unit in the recipient sample and may calculate a "hit rate" therefrom.

- **Second level:**
  After statistical matching the true joint distribution of all variables is reflected in the statistically matched file; i.e., $\widetilde{f}_{X,Y,Z} = f_{X,Y,Z}$.

- **Third level:**
  The correlation structure and higher moments of the variables are preserved after statistical matching with $\widetilde{cov}(X, Y, Z) = cov(X, Y, Z)$. Also the marginal distributions are reflected correctly with $\widetilde{f}_{Y,Z} = f_{Y,Z}$ and $\widetilde{f}_{X,Z} = f_{X,Z}$.

- **Fourth level:**
  After statistical matching, at least, the marginal and joint distributions of the variables in the donor sample are preserved in the statistically matched file. Then $\widetilde{f}_Y = f_Y$ and $\widetilde{f}_{Y,Z} = f_{Y,Z}$ are expected to hold if Y is imputed in the (X, Z) sample.

Any discussion of validity of a statistical matching technique hereinafter is based on these four levels. In the following sections they are discussed in more detail.

2.5.1 *First Level: Preserving Individual Values*

The first and most challenging level of the matching objectives seems to be the reconstruction of the individual values. Since we do not know the true values any hit rate can only be calculated by means of simulation studies. In general, the reproduction of the exact values will only be possible if the common variables Z already determine the variable Y exactly. Such a functional dependence for all units of both samples could be, for example, $Y_i = \alpha + \beta Z_i$, or even $Y_i = Z_i^2$, $i = 1, 2, \ldots, n$. In this case the imputed y value is the "true" one for any given value $Z = z$. Usually this will not happen, especially when the Y variables are continuous and have a multivariate structure. Within continuous distributions the probability of drawing a certain value y is zero; counting the hits is meaningless then. In the case of discrete or classified variables Y a hit rate may be calculated for the purpose of demonstration, counting a hit for the imputation of a p-dimensional variable Y when the whole imputed vector equates the original vector,; i.e.,

$$(\widetilde{y}_{1i}, \widetilde{y}_{2i}, \ldots, \widetilde{y}_{pi}) = (y_{1i}, y_{2i}, \ldots, y_{pi})$$

for $i = 1, 2, \ldots, n_A$. Notice that the calculation of a single hit rate for each variable may mislead the interpretation because it does not ensure that the joint distributions are well preserved.

2.5.2 Second Level: Preserving Joint Distributions

We have assumed that the units of both samples are drawn independently within and between the two samples and the statistically matched file can be regarded as a random sample from the underlying fusion distribution $\widetilde{f}_{X,Y,Z}$. The most important objective of statistical matching is the generation of a complete sample that can be used as a single-source sample drawn from the underlying distribution $f_{X,Y,Z}$: it is less the reconstruction of individual values but the possibility of making any valid statistical analysis based on the statistically matched file. We have shown in Section 2.3.2 that this is only possible if the specific variables Y and X are conditionally independent given the common variables $Z = z$; i.e., $f_{X,Y|Z} = f_{X|Z} f_{Y|Z} = \widetilde{f}_{X,Y|Z}$ holds.

2.5.3 Third Level: Preserving Correlation Structures

Sometimes the analyst's interests are more specific concerning, for instance, only the association of variables measured by their correlation structure. Then the statistically matched file must be considered as randomly generated from an artificial population which has, at least, the same moments and correlation structure as the actual population of interest. From (2.9) we have seen that

$$cov(X, Y) = E(cov(X, Y|Z)) + cov(E(X|Z), E(Y|Z))$$

and $\widetilde{cov}(X, Y) = cov(E(X|Z), E(Y|Z))$ hold, but only $E(X|Z = z)$ and $E(Y|Z = z)$ can be derived theoretically from the separate samples. Thus, the fusion covariance $\widetilde{cov}(X, Y)$ only equals the true $cov(X, Y)$, if X and Y are on the average conditionally uncorrelated given $Z = z$; i.e., $E(cov(X, Y|Z)) = 0$. Notice that variables which are conditionally independent are also conditionally uncorrelated and, of course, on the average conditionally uncorrelated, but not vice versa in general.

2.5.4 Fourth Level: Preserving Marginal Distributions

A minimum requirement of a statistical matching procedure should be the preservation of the distributions that can already be observed in the separate samples. Analysis concerning the marginal distributions based on the statistically matched file should provide the same valid inference when based on the separate samples.

In practice the empirical distributions of the common variables Z as well as the imputed variables Y in the resulting statistically matched file can be compared with their empirical distributions in the donor sample to evaluate the similarity of both samples. The empirical distributions $\widehat{\widehat{f}}_Y$ and $\widehat{\widehat{f}}_{Y,Z}$ should not differ from $\widehat{f}_Y$ and $\widehat{f}_{Y,Z}$ more than two random samples drawn from the true underlying population. Notice that this implies the different samples being drawn according to the same sampling design. Alternatively, weights have to be applied accounting for differing selection probabilities of the units in the separate samples.

2.5.5 Discussion

The first level is obviously very difficult to achieve and in many cases not practical. The second level is the most interesting one because it ensures that any analysis can be done referring to the statistically matched file as a real single-source sample. An efficient statistical matching technique preserves the joint distributions and allows us to make any individual inference based on the statistically matched file. The third level is sufficient, if analysis is restricted (e.g., to the correlation structure). Since the matched data are often intended to become public-use files (i.e., after statistical matching the artificial sample is used by different agencies and analysts for their individual purposes), conditional independence of the variables not jointly observed should be ensured.

The last level seems to be self-evident. Nevertheless, with high-dimensional data structures and complex survey designs it is quite difficult to achieve even this level of validity by statistical matching. However, it is only testable whether the empirical distributions of the donor sample are like the empirical distributions of the statistically matched file. The achievement of other objectives can either be investigated by means of experimental situations, for example, by performing so-called split-half experiments and other simulation studies, or by using a third complete data source. In the typical matching situation only the fourth level can be controlled. Therefore, often statistical matching is said to be successful if the marginal and joint empirical distributions of Z and Y as they are observed in the donor sample are "nearly" the same in the statistically matched file; see Wendt (1976, 1979), Baker et al. (1989), and Koschnick (1995).

The German association for media analysis ("Media Analysis Association" called in German Arbeitsgemeinschaft Media Analyse, for short, AG.MA[2]), for example, postulates the following data controls after a match has been performed.

---

[2]The AG.MA is a media association, i.e., publishing houses, radio and TV stations, and many advertising agencies, as well as a certain number of advertisers.

- First the empirical distributions of the common variables Z in the recipient and the donor sample are compared to evaluate whether their marginal distributions are the same in both samples.
- Next the empirical distributions of the imputed variables Y in the recipient and the donor sample are compared.
- Finally the joint distribution $f_{Z,Y}$ as observed in the donor sample is compared to the joint distribution $\widetilde{f}_{Z,Y}$ as observed in the statistically matched file.

All these comparisons are done using different tests such as χ^2-tests or t-tests to compare empirical distributions or their moments. A successful match should lead to similar relationships between common and specific variables in the donor and the statistically matched file; discrepancies should not be larger than expected between two independent random samples from the same underlying population. In particular, often each pair of variables Y and Z in the donor sample is simply tested at a significance level α for positive or negative association by, for example, a χ^2-test or a t-test (depending on the scale of the variables). Then the same test of association between Y and Z is performed for each pair in the statistically matched file. If the results of the tests only differ in about α percent of the possible (Y, Z) combinations, then the matching procedure is regarded as successful. Among others, nonparametric tests and multiple regression models may be used in the same manner.[3]

However, all these tests may only indicate the fourth-level validity by a matching technique. Without a small but completely observed data set no validation of the other levels is possible. Further evaluations are left to simulation studies.

2.6 Some Illustrations

2.6.1 Conditional Independence

As we have seen, the validity of the matching process depends on the ability of the linking variables to describe the association between, for example, media exposure and purchasing behavior. Given these common variables the variables never jointly observed are more or less independent after statistical matching.

First we give a concrete example for the reader unfamiliar with the meaning of conditional independence. For demonstration purposes we consider a simple situation about the purchases of garlic pills. Raetzel (2000), pp.

[3]For details of these control mechanisms see Koschnick (1995), pp. 674–676. In Chapter 5 more tests and comparisons are discussed and applied to media data.

True relationship of consuming and advertising behavior

Purchasing garlic pills	Children		Elderly people		Sum
	Advertising: yes	Advertising: no	Advertising: yes	Advertising: no	
Yes	5	0	50	10	65
No	95	100	50	90	335
Sum	100	100	100	100	400

Consumer panel

Purchasing garlic pills	Children	Elderly people	Sum
Yes	5	60	65
No	195	140	335
Sum	200	200	400

Television panel

Advertising	Children	Elderly people	Sum
Yes	100	100	200
No	100	100	200
Sum	200	200	400

Mean artificial fusion sample derived from 1000 simulations

Purchasing garlic pills	Children		Elderly people		Sum
	Advertising: yes	Advertising: no	Advertising: yes	Advertising: no	
Yes	2.49	2.51	29.92	30.03	64.95
No	97.51	97.49	70.08	69.97	335.05
Sum	100.00	100.00	100.00	100.00	400.00

Figure 2.4. Hypothetical example and simulation

71–73, performed a simulation study to illustrate this. Let us assume a hypothetical joint distribution for the purchases of garlic pills according to age and advertisement as displayed in Figure 2.4. Basically in this example it is assumed that buyers of garlic pills are usually elderly people. Now let the advertisement for garlic pills just be scattered around randomly at any TV time. We would expect elderly people as well as some children having seen the advertisement about garlic pills to buy them more often.

1000 data sets of size 400 each are generated according to the marginal distributions and linked by age to create new complete data sets. The last table of Figure 2.4 shows that conditional independence is established in the matched data set; i.e., purchasing and viewing the advertisement are independent given the age. Within an age group the purchases of garlic pills are now nearly the same, whether the advertisement has been seen or not.

This small example may illustrate the conditions we need to perform a successful match. In the case of random samples this assumption is conditional independence. The association between the variables X and Y never jointly observed must be explained by the common variables Z. All information about the association between X and Y is already incorporated in Z. If Z is fixed nothing is left to learn from Y about X. Applied to media and consuming data the principle of conditional independence demands that the association between, e.g., media exposure and consuming behavior is completely determined by the common variables. In Section 1.4 the possible range of the unconditional association of X and Y is discussed when the common variables do not exactly determine the specific variables. Within media and consuming data the typical demographic and socioeconomic variables will surely not completely explain media exposure and consuming behavior. Variables already concerning media exposure and consuming behavior have to be asked as well. Thus, the common variables also have to contain variables concerning television and consuming behaviors. In German market research, for example, participants in a television measurement panel are asked additionally about their purchases during some time period. Participants in a consumer panel are asked some questions about their TV viewing behavior. Clearly, many difficulties arise, for instance, in making the purchases as remembered by the television measurement panel participants comparable to the scanned purchases of the consumer panel participants. However, Roberts (1994) reports better results using such "specific common" variables than the usual demographic and socioeconomic issues alone.

2.6.2 Assuming Normally Distributed Data

Under the assumption of a particular data model further derivations are possible. Let us assume an r-multivariate normal distribution for the joint distribution of X, Y, and Z; i.e., $(X, Y, Z) \sim N_r(0, \Sigma)$ with zero means and the $r \times r$ covariance matrix

$$cov(X, Y, Z) = \Sigma \quad = \quad \begin{pmatrix} \Sigma_{XX} & \Sigma_{XY} & \Sigma_{XZ} \\ \Sigma_{YX} & \Sigma_{YY} & \Sigma_{YZ} \\ \Sigma_{ZX} & \Sigma_{ZY} & \Sigma_{ZZ} \end{pmatrix},$$

$\Sigma_{YX} = \Sigma'_{XY}$, and so on. Notice that Σ_{XX}, Σ_{YY}, and Σ_{ZZ} usually denote the covariance matrices which contain the variances and covariances of each pair of variables. If, for example, X has q dimensions with $X = (X_1, X_2, \ldots, X_q)$, then the covariance matrix concerning X is $\Sigma_{XX} = E(X - E(X))(X - E(X))'$ which, when written out in full, is

$$\Sigma_{XX} = \begin{pmatrix} \sigma_{X_1X_1} & \cdots & \sigma_{X_1X_q} \\ \cdots & \cdots & \cdots \\ \sigma_{X_qX_1} & \cdots & \sigma_{X_qX_q} \end{pmatrix}$$

with $var(X_1) = \sigma_{X_1X_1} = \sigma^2_{X_1}, \ldots, var(X_q) = \sigma_{X_qX_q} = \sigma^2_{X_q}$. Analogously, we usually calculate $\Sigma_{XZ} = E(X - E(X))(Z - E(Z))'$, $\Sigma_{YZ} = E(Y - E(Y))(Z - E(Z))'$, and $\Sigma_{XY} = E(X - E(X))(Y - E(Y))'$ denoting the matrices of the covariances between the variables.

Under the assumption of a trivariate normal distribution for X, Y, and Z, Gabler (1997) shows that

$$\widetilde{cov}(X, Y) = \tilde{\sigma}_{XY} = \frac{\sigma_{XZ}\sigma_{YZ}}{\sigma^2_Z} \tag{2.16}$$

holds. Generalizing this result, we achieve the covariance matrix established by statistical matching if a multivariate normal data model is assumed with

$$\widetilde{cov}(X,Y) = \widetilde{\Sigma}_{XY} = \Sigma_{XZ}\Sigma_{ZZ}^{-1}\Sigma_{ZY}. \tag{2.17}$$

Using expression (2.11) this leads to

$$E(cov(X,Y|Z)) = cov(X,Y) - \widetilde{cov}(X,Y) = \Sigma_{XY} - \Sigma_{XZ}\Sigma_{ZZ}^{-1}\Sigma_{ZY}.$$

By (2.16) and (2.17) the result of the match concerning the unconditional covariance of X and Y is described whatever normal model may have generated the data. Hence, we find that after the match the variables X and Y are computed uncorrelated without respect to their initial correlation, if X, Z or Y, Z are uncorrelated. Otherwise, if X, Z and Y, Z are correlated, then X, Y are computed correlated as well, although they may be uncorrelated initially.

2.6.3 Simulation Study

The theoretical framework derived so far is based on the assumption that exact matches in terms of the common variables are made. In practice we will observe many common variables and an exact match on them is not realistic at all. Thus, we have already introduced some nearest neighbor matching algorithms as well as the propensity score matching procedure. So there are two objectives of this simulation study. We want to investigate the influence of a nearest neighbor and propensity score match instead of an exact match. On the other hand, the assumption of conditional independence is violated and its effect on the algorithms investigated. The design of the simulation study is kept simple for illustration purposes and is used again at the end of Chapter 4.

Let (Z_1, Z_2, X, Y) each be univariate standard normally distributed variables. Let $(Z_1, Z_2, X, Y) \sim N_4(0, \Sigma)$ with covariance

$$\Sigma = \left(\begin{array}{cc|c|c} 1.0 & 0.2 & 0.5 & 0.8 \\ 0.2 & 1.0 & 0.5 & 0.6 \\ \hline 0.5 & 0.5 & 1.0 & 0.8 \\ \hline 0.8 & 0.6 & 0.8 & 1.0 \end{array}\right) = \begin{pmatrix} \Sigma_{ZZ} & \Sigma_{ZX} & \Sigma_{ZY} \\ \Sigma_{XZ} & \sigma_{XX} & \sigma_{XY} \\ \Sigma_{YZ} & \sigma_{YX} & \sigma_{YY} \end{pmatrix}. \tag{2.18}$$

We assume that the true covariance between X and Y is given with $\sigma_{XY} = \sigma_{YX} = 0.8$. Then the true conditional correlation between X and Y given $Z = z$ is

$$\begin{aligned} \rho_{XY|Z} &= \frac{\sigma_{XY|Z}}{\sqrt{\sigma_{XX|Z}\sigma_{YY|Z}}} \\ &= \frac{\sigma_{XY} - \Sigma_{XZ}\Sigma_{ZZ}^{-1}\Sigma_{ZY}}{\sqrt{(\sigma_{XX} - \Sigma_{XZ}\Sigma_{ZZ}^{-1}\Sigma_{ZX})(\sigma_{YY} - \Sigma_{YZ}\Sigma_{ZZ}^{-1}\Sigma_{ZY})}} \qquad (2.19) \\ &= \frac{0.8 - 0.5833}{\sqrt{(1 - 0.4167)\cdot(1 - 0.8417)}} = \frac{0.2167}{\sqrt{0.5833 \cdot 0.1583}} = 0.7129. \end{aligned}$$

Furthermore, let the recipient sample be (Z_1, Z_2, X) and the donor sample be (Z_1, Z_2, Y); thus X and Y are never jointly observed and Y is imputed in the recipient sample.

Conditional independence of $X, Y|Z = z$ leads to the unconditional covariance after statistical matching

$$\widetilde{\sigma}_{XY} = \Sigma_{XZ}\Sigma_{ZZ}^{-1}\Sigma_{ZY} = (0.5\ 0.5)\begin{pmatrix} 1 & 0.2 \\ 0.2 & 1 \end{pmatrix}^{-1}\begin{pmatrix} 0.8 \\ 0.6 \end{pmatrix} = 0.5833.$$

We draw $n = 5000$ random numbers for (z, x, y) according to $(Z, X, Y) \sim N_4(0, \Sigma)$ as given with (2.18). Then this generated data set is divided randomly into two parts, each sample now of size 2500 and all X in the donor sample or Y values in the recipient sample are eliminated. Then we use different matching techniques to match the two samples; in particular these are as follows.

- Polygamy (nn): a nearest neighbor match based on the minimum absolute distance of the Z values with no restriction on the multiple use of donor units.
- Bigamy (nn): a nearest neighbor match based on the minimum absolute distance of the Z values with donor units only allowed to be used twice.
- Triple mean: the mean of the three nearest neighbor units based on the minimum absolute distance of the Z values is imputed.
- Polygamy (ps): a propensity score matching is performed based on the minimum absolute distance of the estimated propensity score $\widehat{e}(z)$ with no restriction on the multiple use of donor units.
- Bigamy (ps): finally again the propensity score matching is performed on the basis of the minimum absolute distance of the estimated propensity score $\widehat{e}(z)$ but each donor may only be used twice.

In the first three cases the distance function is $d(z_i, z_j) = |z_{1i} - z_{1j}| + |z_{2i} - z_{2j}|$ for $i = 1, 2, \ldots, n_A$, $j = 1, 2, \ldots, n_B$, $n_A = n_B = 2500$. In the last two cases we use $d(z_i, z_j) = |\widehat{e}(z_i) - \widehat{e}(z_j)|$ for $i = 1, 2, \ldots, n_A$, $j = 1, 2, \ldots, n_B$, $n_A = n_B = 2500$.

Technique	$\widehat{E}(\widehat{\mu}_Y)$	$\widehat{E}(\widehat{\sigma}_Y^2)$	$\widehat{E}(\widehat{\sigma}_{Z_1Y})$	$\widehat{E}(\widehat{\sigma}_{Z_2Y})$	$\widehat{E}(\widehat{\sigma}_{XY})$	$\widehat{E}(\widehat{\rho}_{XY})$	$\widehat{E}(\widehat{\rho}_{XY\|Z})$
Polygamy (nn)	-0.0009 (0.0223)	0.9939 (0.0327)	0.7949 (0.0331)	0.5953 (0.0196)	0.5792 (0.0250)	0.5801 (0.0135)	-0.0006 (0.0061)
Bigamy (nn)	-0.0012 (0.0212)	0.9928 (0.0312)	0.7944 (0.0318)	0.5942 (0.0203)	0.5776 (0.0250)	0.5788 (0.0132)	-0.0015 (0.0057)
Triple mean	-0.0008 (0.0218)	0.8755 (0.0288)	0.7899 (0.0312)	0.5909 (0.0198)	0.5749 (0.0252)	0.6135 (0.0132)	-0.0010 (0.0040)
Polygamy (ps)	-0.0026 (0.0296)	0.9996 (0.0400)	0.4149 (0.3010)	0.3053 (0.3355)	0.3007 (0.1870)	0.3021 (0.1887)	0.0016 (0.0103)
Bigamy (ps)	-0.0023 (0.0249)	0.9975 (0.0357)	0.4146 (0.2989)	0.3045 (0.3364)	0.3002 (0.1869)	0.3019 (0.1889)	0.0016 (0.0100)

Table 2.1. Simulation study with nearest neighbor (nn) and propensity score (ps) matches, $\rho_{XY|Z} \neq 0$

The whole procedure of generating and dividing the data, matching the two samples as described above, and calculating the usual estimates of mean, variance, and covariances is repeated $k = 50$ times. From these empirical distributions the expectations $E(\widehat{\mu}_Y)$, $E(\widehat{\sigma}_Y^2)$, $E(\widehat{\sigma}_{ZY})$, $E(\widehat{\sigma}_{XY})$, $E(\widehat{\rho}_{XY})$, and $E(\widehat{\rho}_{XY|Z})$ are estimated by simply averaging over $k = 50$ repetitions; i.e., $\widehat{E}(\widehat{\theta}) = (1/k)\sum_{j=1}^{k} \widehat{\theta}_k$. Also the sample standard error of each estimate $\widehat{\theta}$ is computed with

$$s(\widehat{\theta}) = \sqrt{\frac{1}{k-1}\sum_{j=1}^{k}(\widehat{\theta}_j - \widehat{E}(\widehat{\theta}))^2} \qquad k = 50\,.$$

The results are presented in Table 2.1; the sample standard errors $s(\widehat{\theta})$ are given in brackets. Table 2.1 shows that the nearest neighbor algorithms reproduce the moments of the marginal distributions quite well. Furthermore, conditional independence is established by every matching technique. Although the true covariance between X and Y is 0.8, after statistical matching values near 0.58 are calculated. The conditional correlation of X and Y given $Z = z$ is computed according to (2.19) using the ML variance and covariance estimates of the elements of Σ. In each simulation run we find $\widehat{\rho}_{XY|Z} \approx 0$, although the true conditional correlation is $\rho_{XY|Z} = 0.7129$. The sample standard error of $\widehat{\rho}_{XY|Z}$ is very small.

Matching by means of propensity scores also produces conditional independence but alters the joint distribution of (Y, Z). Exploring the estimated scores $\widehat{e}(z)$, we find that usually values of $\widehat{e}(z) \approx 0.5$ are calculated, because there is no difference between the two groups. They are generated as a random sample from the underlying population; thus the Z variables do not separate the groups here. The association between Y and Z is underestimated with approximately half of their true values after statistical matching. Thus, the fusion covariance of propensity score matching is estimated with $\widehat{\widehat{\sigma}}_{XY} = \widehat{\Sigma}_{XZ}\widehat{\Sigma}_{ZZ}^{-1}\widehat{\Sigma}_{ZY} \approx 0.3$. The variabilities of the covariance estimates are rather high then.

Another simulation run is performed with the covariance between X and Y set to 0.6; thus, more or less conditional independence is assumed with $\rho_{XY|Z} = 0.0550$. This leads to the same conclusions; the results of the estimated expectations are presented in Table 2.2 with the sample standard errors given in parentheses.

Again we find the variance of Y best preserved by propensity score matching, although the covariance between Y and Z is even more distorted then. And again conditional independence is established by every matching technique. Finally, the parameters reproduced by statistical matching are not affected by merging nearest neighbor units instead of exact statistical matches. The influence of different marriage processes on the reproduced parameters is likewise low.

Technique	$\check{E}(\widehat{\mu}_Y)$	$\check{E}(\widehat{\sigma}^2_Y)$	$\check{E}(\widehat{\sigma}_{Z_1Y})$	$\check{E}(\widehat{\sigma}_{Z_2Y})$	$\check{E}(\widehat{\sigma}_{XY})$	$\check{E}(\widehat{\rho}_{XY})$	$\check{E}(\widehat{\rho}_{XY\|Z})$
Polygamy (nn)	0.0019 (0.0209)	0.9883 (0.0411)	0.7921 (0.0349)	0.5960 (0.0265)	0.5831 (0.0269)	0.5853 (0.0138)	0.0010 (0.0073)
Bigamy (nn)	0.0019 (0.0199)	0.9864 (0.0363)	0.7912 (0.0335)	0.5940 (0.0255)	0.5822 (0.0267)	0.5849 (0.0144)	0.0013 (0.0065)
Triple mean	0.0022 (0.0194)	0.8745 (0.0339)	0.7891 (0.0328)	0.5924 (0.0240)	0.5794 (0.0253)	0.6183 (0.0137)	0.0001 (0.0034)
Polygamy (ps)	-0.0038 (0.0278)	0.9977 (0.0320)	0.3459 (0.3019)	0.2642 (0.3406)	0.2546 (0.1986)	0.2552 (0.1999)	0.0011 (0.0141)
Bigamy (ps)	-0.0006 (0.0236)	0.9995 (0.0335)	0.3438 (0.3023)	0.2634 (0.3406)	0.2528 (0.2009)	0.2535 (0.2021)	0.0004 (0.0137)

Table 2.2. Simulation study with nearest neighbor (nn) and propensity score (ps) matches, $\rho_{XY|Z} \approx 0$

Care has to be taken, when the mean of, for instance, three nearest neighbor units is imputed (i.e., $\widetilde{y}_i = \overline{y}_i$ for $i = 1, 2, \ldots, n_A$), then the imputed variable obviously has less variability than the original one. In the case of multivariate normally distributed Y and Z variables, the variance $\widetilde{var}(Y)$ of any mean of k nearest neighbor observations Y after statistical matching is generally given by

$$\begin{aligned}
\widetilde{var}(Y) &= E(\widetilde{var}(Y|Z)) + var(\widetilde{E}(Y|Z)) \\
&= E(\underbrace{var(\overline{Y}|Z)}_{= var(Y|Z)/k}) + var(\underbrace{E(\overline{Y}|Z)}_{= E(Y|Z)}) \\
&= E\left(\Sigma_{YY} - \Sigma_{YZ}\Sigma_{ZZ}^{-1}\Sigma_{ZY}\right)/k + var\left(\mu_Y + (Z - \mu_Z)\Sigma_{ZZ}^{-1}\Sigma_{ZY}\right) \\
&= \frac{1}{k}(\Sigma_{YY} - \Sigma_{YZ}\Sigma_{ZZ}^{-1}\Sigma_{ZY}) + \Sigma'_{ZY}\Sigma_{ZZ}^{-1}var(Z)\Sigma_{ZZ}^{-1}\Sigma_{ZY} \\
&= \frac{1}{k}\Sigma_{YY} + (1 - \frac{1}{k})\Sigma_{YZ}\Sigma_{ZZ}^{-1}\Sigma_{ZY},
\end{aligned}$$

making use of the parameters of a conditional normal density; see Mood et al. (1974), p. 167. In our simulation study the value of the theoretical fusion variance of the triple mean is $\widetilde{var}(Y) = \frac{1}{3} + \frac{2}{3}0.8417 = 0.8944$ which is slightly underestimated by the simulations. To ease the interpretation we use a t-statistic-like value according to $\sqrt{k}(\widehat{E}(\widehat{\theta}) - \widetilde{\theta})/s(\widehat{\theta})$. Calculating these t-statistics we never get an absolute value greater than 3 for all the estimated expectations derived from the nearest neighbor matches polygamy and bigamy. With the triple mean the theoretical fusion variance $\widetilde{var}(Y)$ as well as the covariances tend to be underestimated; thus the unconditional correlation of X and Y is inflated. With propensity score matching only the mean and the variance of Y and the conditional fusion correlation of zero are reproduced well. All other estimates are obviously biased and have a rather large sample variability.

However, it can be difficult to calculate the exact formulas of the covariance reproduced by the matching algorithm as is done with (2.16) or (2.17). Therefore another simulation study was performed by Rässler and Fleischer (1998) to investigate if, and to what degree of validity, the presented results hold for different data models and different marriage algorithms. There again the nearest neighbor matches with and without penalty function produce the fusion distribution $\widetilde{f}_{X,Y,Z}$ and conditional independence quite well. Matching of data sets using algorithms like nearest neighbor matches can reproduce the true correlation between variables X and Y not jointly observed only if they are uncorrelated on the average conditional on the common variable $Z = z$, i.e., if $E(cov(X, Y|Z)) = 0$ holds. The stronger demand for conditional independence is not necessary if the interest is focused on the correlation (or higher moments) between X and Y only.

Other simulation studies concerning departures from the conditional independence assumption and the use of auxiliary information were published by Rodgers (1984), Paass (1986), Singh et al. (1993), and Liu and Kovacevic (1994, 1998). They all lay stress upon the conditional independence assumption and compare the validity of different matching techniques concerning the second and third levels. Little evidence is found that suggests using any version of, for example, the Mahalanobis distance function instead of subjectively weighted sums of the absolute differences between the Z variables; see Liu and Kovacevic (1998). In general, all procedures investigated in simulation studies seem to establish more or less conditional independence whatever the true relationship of the variables X and Y never jointly observed is.

2.7 Concluding Remarks

A frequentist framework is derived herein based on exact matches of the common variables Z. We have shown that statistical matching techniques based on the matching of such statistical twins will lead to conditional independence between the variables never jointly observed, i.e., X and Y given $Z = z$. All marginal distributions of (X, Z) and (Y, Z) are preserved well in this framework. In simulation studies no greater differences from the conditional independence assumption are found when nearest neighbor algorithms are used instead of exact matches. Also different marriage algorithms and distance functions do not affect the derived results considerably; only the imputation of the mean of k nearest neighbor units reduces the variance of the imputed variable. The extent to which the joint distributions are distorted by matching via balancing scores is discussed in a simulation study. The closer the match is to the exact match in the common variables, that is, the finer the balancing score, the less distortion is observed in the joint distributions of X, Z or Y, Z after statistical matching.

Structuring the objectives of statistical matching, we argue that the most important target is the preservation of the unconditional joint distributions. In an ideal situation any analysis based on the statistically matched file may be performed regarding the matched file as a random sample from the underlying population. Especially in European media planning this is the matching task. However, usually the common variables will not determine the conditional association of the variables not jointly observed completely. Suitable variables to be used as common variables must be searched in a large data set although the assumption of conditional independence is not yet testable by the observed data. The separate samples contain no information about the appropriateness of this assumption and information from other sources must be considered. Thus, a small but completely observed data set is necessary to test whether variables may serve as common variables Z (i.e., whether they may establish conditional independence). Clearly, with high-dimensional data this again is a difficult task. The simulation studies indicate that the choice of suitable common variables is more important concerning the validity of a statistical matching technique than the matching technique itself. Therefore, many matching strategies applied in practice make use of quite sophisticated procedures to find or even construct suitable common variables. This is discussed in the following chapter.

3

Practical Applications of Statistical Matching

A theoretical mathematician and a research statistician were each situated in opposite corners of a square room. In the corner between the two men was a lovely young lady waiting to offer her charms to whichever of them reached her first. Each man had to proceed towards the young lady one step at a time. And each step taken could only cover half the remaining distance between each man and the young woman.
Upon hearing the rules, the theoretical mathematician left the chamber because he realized that it was impossible ever to get to the lovely young damsel. The research statistician stayed and claimed the prize. While he too realized it was impossible to reach her, he figured that he could get close enough for all practical purposes.

Attributed to Leslie Kish according to Okner (1972b).

3.1 Introduction and Chapters Outline

Much of the literature describing traditional approaches and techniques that are used in practice are working papers, technical or internal reports. Often they are difficult to obtain if available at all. Most of the reports or articles published are less theoretical; details about the final matching algorithms are often best explained in private talks or at conferences. No comprehensive work addressing new and recently used matching techniques is available. In this chapter we summarize and record the history of statistical matching techniques and briefly explain some of its first solutions.

Different techniques that had and still have great importance for practical applications are then discussed in more detail. Often our information is based on unpublished reports supplied by experts practicing statistical matching. Thus we hope to fill a gap in the literature and explain what often is left to the reader's imagination.

Since statistical matching has developed from quite different points of view mainly in Europe as well as the United States and Canada we start with describing its growth. Then an overview is given and the matching techniques as they are currently applied to practical problems are presented. We go into detail as far as necessary to illustrate the main ideas. The algorithms themselves are not described explicitly if detailed literature is available.

As shown in Chapter 2 these matching techniques may produce good results when the assumption of conditional independence is valid. Therefore, the procedures are finally discussed from the conditional independence viewpoint. We conclude that the traditional approaches put much effort into carefully deriving suitable common variables and into preserving at least the fourth level of validity. However, we do not know whether the common variables may dissolve the association between the variables not jointly observed. We do not even know how much of the association can be explained by the common variables. Thus, we want to draw the reader's attention to the alternative approaches presented in Chapter 4.

3.2 History of Statistical Matching Techniques

3.2.1 Data Fusion in Europe

According to Wendt (1976) and Koschnick (1995), pp. 661–662, the first statistical matching techniques were applied independently from another between 1963 and 1965 in Germany, France, and the United Kingdom. In the UK Beale used a linear programming technique to match television and media data. Boucharenc and Bergonier matched data in France by means of segmentation, distance functions, and a randomized order of assignment. Their procedure seems to be very close to the later solution widely used in Europe. In Germany Wendt developed a simulation program based on segmentation as well as random assignment. These first approaches were not very successful obviously due to many technical restrictions. Some years later the French and the German procedures were combined and practical applications arose; see Bergonier et al. (1967), Bergonier (1968), and Wendt and Wendt-Maeder (1979), p. 54. Following Baker et al. (1989), in the UK the research industry was quite reluctant to perform statistical matching, although it became more and more common in France and Germany.

In Germany statistical matching is connected with the German Media Analysis Association, the AG.MA,[1] and the Bureau Wendt. The AG.MA was founded in 1954 as "Arbeitsgemeinschaft Leseranalyse" (AGLA) (i.e., "Readership Analysis Association") and reorganized as the Media Analysis Association in 1971; at the beginning a survey of magazine readership was conducted every two years. Soon this readership survey was extended to any sort of print media; radio and television consumption was also brought in. A study has been published every year since 1960. The demand for more information describing all sorts of media target groups has grown steadily over the years. A nearly all-media survey arose; it is still called Media Analysis (MA) and has become a basic survey, essential for intra- and intermedia planning in Germany. The steady growth of consumer journals, the introduction of new TV and radio programs, and the emergence of new media soon brought Media Analysis to its limits. To survey the growing market and still meet the intermedia data needs of media planning the single source principle finally was abandoned. A suitable tool for combining sets of data gathered separately was searched ensuring an "optimal" balance between survey requirements and computational advantages. Thus, methods of marrying data, such as matching procedures, came into consideration to create large databases to support all kinds of media planning. Bureau Wendt played an important role during these years and developed a matching technique, the so-called "topological concept". Beginning in the later 1970s, AG.MA began and allowed regularly merging the MA data with data from separate sources according to the rules of the so-called "German partnership model".[2] After matching, the data are used to estimate audience coverage and to develop media schedules based thereon. Statistical matching is judged to be successful if results do not deviate more than expected from results obtained from independent samples. This ensures the appropriateness of the fourth-level validity according to our terminology as defined in Section 2.5. At the end of the 1960s and during the 1970s different matching techniques were tested, exposés written and improvements made; see, e.g., Wendt and Wendt-Maeder (1986), pp. 333–391. Wendt (1975) gives a comparison of typology systems used in German market research at that time. Problems and possibilities of matching techniques using segmentation and typological approaches were discussed by Wendt-Maeder (1976) allowing a first insight to a wider public.

---

[1] See Section 2.5.5.

[2] Administrative as well as technical rules are formulated there; the first version dates back to 1978 and a new and extended version was published in 1996. If data from the MA are matched with data from different sources observation of these rules including given test strategies are demanded; see Arbeitsgemeinschaft Media-Analyse e.V. (1996), Scheler (1993), and Wiegand (1986). The history of the AG.MA and the development of the German partnership model are described in great detail by Wendt and Wendt-Maeder (1979, 1983, 1986).

The Worldwide Readership Research Symposium began in New Orleans in 1981 and since then it has taken place every two years in different locations around the world: Montreal followed in 1983 and Salzburg in 1985. New procedures to improve readership measurement are discussed there and regular sessions deal with statistical matching techniques. These first symposia seem to have had a great influence on the development and distribution of statistical matching techniques mainly in Europe. Thus, similar techniques were established focusing on the search of typologies or some nearest neighbor algorithms based on a variety of distance functions. In Germany extensive experiments were conducted during the 1980s (see Wiegand (1986)) by dividing the 1984 MA into two parts, matching these parts again, and evaluating the results. In these experiments the topological concept was found to be the best of the methods tested by AG.MA. The tests compared the empirical marginal distributions of the imputed variable in the recipient sample before and after the match; thus the validity measured concerned the fourth level only. No tests were reported concerning the preservation of, at least, the association of the artificially split variables; see Scheler and Wiegand (1987). Finally, the topological concept has been continuously improved by Wendt (1975, 1976, 1979, 1986) and it became the basis of the German partnership model of the AG.MA. Notice that the main idea of the German partnership model is not to bring together as many surveys as possible forming one big statistically matched file, but to perform all the surveys in such a manner that they may serve as donor surveys especially for the MA. Moreover, rules are formulated based on the topological concept ensuring general acceptance of a statistically matched file, if the matching procedure is done accordingly; see Koschnick (1995), pp. 673–676.

During the 1980s in Europe two algorithms seemed to be most widely used. The first was the German partnership model based on the topological concept and the second one is called the FRF (Fusion sur Réferenciel Factoriel) algorithm, designed by Santini and based on a hierarchy of marriages where the transferred information is possibly given from more than one donor (see Antoine and Santini (1987)). The French system worked on the basis of a χ^2-metric distance function between the recipient and the donor unit. This distance function is the basis of a correspondence analysis that maps the units from the recipient and the donor file into a multi-dimensional space. New distance measures are then calculated based on the coordinates of the mapped recipient and donor unit. Finally, different types of marriages are defined and optimization algorithms are used to achieve the final matches. An actual overview of recent matching techniques discussed in the French literature is given by Aluja-Banet and Thio (2001). They describe, for example, statistical matching based on k-nearest neighbor observations or using principal component regression analysis.

In the later 1980s European countries including the UK, Belgium, Finland, and Spain started matching experiments and projects; see Antoine and Santini (1987). O'Brien (1991) stated that the first commercially available statistically matched file in the UK was based on the TGI (Target Group Index, a product and media survey) data matched with BARB (Broadcasters' Audience Research Board collecting viewing data) data. Nevertheless, it was not the first statistical matching to be conducted in the UK. According to O'Brien (1991) the match of the National Readership Survey and the Financial Research Survey was one of the first British activities using the French approach developed by Santini. Obviously it was discussed widely in the later 1980s but no file for public use was created. Moreover, the validity of the Santini approach was explored by extensive experiments. Baker et al. (1989) published the results of a major experiment concerning statistical matching that was undertaken using the British TGI data to examine the French matching procedure from Santini. For the forthcoming match of TGI data and the BARB data an algorithm based upon the Mahalanobis distance measurement was applied; see Baker (1990). The importance of the common variables forming a link between the specific variables of the donor and the recipient sample was emphasized for conducting an efficient statistical match. In the TGI/BARB fusion only 11 common variables were available directly and two more common variables were created additionally concerning the total television and commercial television weight of viewing. For the TGI survey this was derived by asking the participants about their viewing behavior, within the BARB survey the viewing was already metered. Experiments and procedures concerning this approach were published in detail by Baker (1990) and O'Brien (1991). Roberts (1994) gave a comprehensive description of matching the BARB data with the so-called AGB Superpanel, a large market tracking panel in the UK. Again stress was laid on the quality of the common variables to serve as good predictors of the behavioral information which was to be transferred. Therefore the Superpanel members were asked about their viewing habits in a small additional questionnaire. These variables play a central role in the matching process, namely, the claimed viewing time by day part, and the programs remembered are used for the matching procedure (see Roberts (1994)). Beginning at the end of the 1990s and initiated by British matching techniques the German market research company GfK developed their own matching technique by introducing such "specific common" variables. Currently the GfK asks participants of its television measurement panel some questions about their consuming behavior. Also participants of the consumer panel are asked a few questions concerning their viewing behavior. This additional information is very important for the matching procedure. Clearly, the claimed viewing habits differ from the more precise metered information and purchases are questioned only on a very aggregated level. However, Baker (1990), O'Brien (1991), and Roberts (1994) report quite satisfying results about their matching procedures. Again most of the evaluations of the matching processes concern the preservation of the marginal distributions.

Notice that the matching of databases is widely accepted to be not necessarily the best solution for media planning, but to be the only practical solution as long as single source information cannot be obtained; see Antoine and Santini (1987), Buck (1989), and Adamek (1994). We find that the methods actually applied use distance measures combined either with different nearest neighbor or marriage algorithms or cluster analysis procedures. Often an iterative search is used to achieve a (more or less) overall minimum distance although the multiple use of donor units usually is restricted in some way. All the known techniques emphasize the "good" explanatory power of the common variables. Nevertheless, reported tests of the validity of the matching techniques generally focus on the preservation of the marginal distributions, i.e., in our terms on sustaining the fourth level of validity.

3.2.2 Statistical Matching in the USA and Canada

Contrary to the development of statistical matching techniques in Europe due to media planning demands by private media institutes, in the US and Canada federal offices started to merge data from different sources.[3] In addition to technical reports and working papers, the first articles about the creation of such synthetic microdata files in the US were published in the early 1970s, for instance, by Okner (1972a, 1974), Budd (1971), and Ruggles and Ruggles (1974). Alter (1974) provided a comprehensive description of a Canadian project, the statistical matching of the Canadian Survey of Consumer Finances with the Family Expenditure Survey. Due to the increasing computational power available and a growing interest in social problems at the microeconomic level, the demand for large amounts of disaggregated economic and demographic information obviously grew more and more in both, the US and Canada during the 1960s and 1970s. The lack of a consistent and comprehensive set of household income data especially stimulated the construction of a new microanalytic database.

During the 1970s changes took place in the laws and regulations in the US concerning exact matching of records (i.e., record linkage) for statistical and research purposes. Additional restrictions were imposed, especially by the Privacy Act of 1974 and the Tax Reform Act of 1976, if data sets that belonged to a federal agency were matched with data of another federal agency or even with those of other organizations; see Goel and Ramalingam (1989), p. 3. Moreover, analysis of microeconomic models often was and is based on data only available from different sources that are results of random samples from a very large population. When two or more random

[3]For instance, the so-called 1966 MERGE File combines information from the 1967 Survey of Economic Opportunity and the 1966 Tax File. The research project was financed under a grant from the US Office of Economic Opportunity; see Okner (1972a).

samples are involved, the probability of the same individual appearing in both samples usually is extremely small if not zero. Thus, in many situations record linkage of files is either theoretically or legally impossible and statistical matching of the separate files is applied.

According to Ruggles and Ruggles (1974), the 1970 Public Use Sample was matched with the Survey of Economic Opportunity file by cross-classifying the data into 288 cells based on five variables. Within these cells the matches were achieved by sequentially ordering the units using another three variables. Okner (1972a) grouped the data from the Survey of Economic Opportunity and the Tax File into equivalence classes using demographic (e.g., marital status and age) and income information (e.g., income source categories and major income source values). Acceptable matches were defined on the grounds of a distance function calculated from consistency scores derived from various income and property criteria. The final match was done randomly with a probability proportional to some Tax File weights. Alter (1974) based his matching procedure on a multivariate regression approach. First the database was segmented into four bases according to home ownership and family type. Then the common variables Z were taken as the independent variables and each dependent specific variable X and Y was regressed upon Z. From the regression results consistency scores were derived and attached to each common variable according to the explanatory power of the specific variables. Based on these scores a distance function was constructed and different acceptance levels defined. For some variables such as major income source or child status code, a match was made mandatory. The final matching was performed by searching to achieve a minimum distance measure for each database.

Ruggles and Ruggles (1974) provided an extensive description of different matching techniques applied in the US in the early 1970s. In general, the procedures were designed to manage large data sets, for instance, the match of 2 million units of the Public Use Sample with another 2 million units of the Social Security File was discussed. Thus, distance measures that have to be minimized in iterative algorithms as they are common in nearest neighbor algorithms or cluster analysis techniques are of less practical importance than procedures adapting cross-tabulation techniques and hierarchical sorting. The generation of large synthetic statistically matched files by actually minimizing a sum of distances over all potential matches often simply exceeded the computational power available in the 1970s. Furthermore, alternatives to distance-based strategies were discussed which explicitly made use of the conditional independence assumption (see Goel and Ramalingam (1989), pp. 78–85), although they obviously did not have any practical importance.

The traditional strategies have been criticized in the file-matching literature on the grounds that such procedures implicitly make the assumption of conditional independence. Such a critique was put forward especially by Sims (1972a and b) and Rodgers (1984). We show in Chapter 2 that this criticism is most appropriate when an observation (Z, Y) is randomly matched with an observation (Z, X) from among all equivalent (Z, Y) observations, as done by Okner (1972a) and many others. Rodgers (1984) summarized the results of those simulation studies where the validity of different matching techniques was investigated concerning the preservation of the marginal distributions as well as the conditional independence assumption, i.e., achieving the fourth and the second level in our terminology. Maybe due to these results published by Rodgers (1984) statistical matching became less important, especially in the United States (see Judkins (1998)).

Statistics Canada did intensive research on improving matching techniques mainly during the 1990s. Extensive simulation studies using large real data sets were reported by Kovacevic and Liu (1994) and Liu and Kovacevic (1997, 1998). They investigated to what extent the earlier findings based on the synthetic data of Paass (1986) and Singh et al. (1993) held under conditions similar to those in real matching environments. It was examined whether the conditional independence assumption could be successfully overcome by the use of appropriate auxiliary information. Furthermore, categorical constraint matching as extended and proposed by Liu and Kovacevic (1996, 1997) and Liu (1998) was compared with many other matching strategies by Liu and Kovacevic (1998) using the 1986 Public Use Micro File and the 1991 Census for the province of Quebec, Canada. Notice that all of these procedures are of the so-called "hot-deck" type, a sort of nearest neighbor search. A hot-deck imputation procedure usually is defined as one where the value imputed to a recipient record comes from a donor record that satisfies certain criteria, for example, having the minimum distance from the recipient unit or belonging to the same class.

Again matching is aimed at achieving policy relevant analysis of tax and transfer programs, to get information about public health and welfare, or, for instance, about educational attainment. Comprehensive databases are required and constituted from different sources by the Canadian federal statistical office, Statistics Canada. For example, the Social Policy Simulation Database is constructed to support microanalytic modeling by combining data from four major sources. Survey data on family incomes and expenditures from the Canadian Survey of Consumer Finances and the Canadian Family Expenditure Survey are combined with administrative data from the Canadian Personal Income Tax Returns and the Canadian Unemployment Insurance Claim histories; see Liu and Kovacevic (1998).

Contrary to the development in Europe we find larger data sets to be merged based on fewer variables that may serve as common variables or are transferred; see Liu and Kovacevic (1998). Increasing computational power allows operating large data sets and makes use of auxiliary information possible. However, the large number of variables to be handled in European matching tasks is still a hindrance to applying similar procedures, for example, such categorically constrained matching routines using auxiliary files.

3.3 Overview of Traditional Approaches

During the 1980s the two basic types of unconstrained and constrained matching were recognized within the realm of statistical matching; see Rodgers (1984). For the time being we may further classify the methods into two main groups depending on whether they utilize auxiliary information. Within these groups we find methods with and without constraints that are imposed basically on the specific variables to be transferred. The overview starts with the principle of the unconstrained match described by means of a simplified example. Then this example is used to explain the idea of constrained matching. We discuss the use of auxiliary information via the categorically constrained matching technique as it was applied and published by Liu and Kovacevic (1997, 1998) and Liu (1998).

The topological concept based on cluster analysis as developed by Wendt is presented in detail here. For a long time it has been the major matching technique used in Europe, and it is still a very important one. Finally, the British solution doing multiple ascription as published by Roberts (1994) and recently adopted by the German market research company GfK is explained. At the end we hope to have covered the various streams of matching techniques actually applied worldwide. Each description ends with a short appraisal concerning the conditional independence assumption.

Notice that often the data are first classified into K "matching classes" (sometimes also called pockets, blocks, or imputation classes) which are identically defined for the separate files. Data sets only within the same class are allowed to be matched. For example, a natural matching class is given by gender: a female can only be matched with another female. This procedure is also often called segmentation and, strictly speaking, leads to K separately matched data sets. All the distance functions used to find suitable matches are applied at the matching class level. Figure 3.1 pictures a simple unconstrained match within each matching class created by gender.

It should further be noticed that it is not mandatory to use all the common variables Z available for the matching process; for example, see in Figure 3.2.

Often the distance function leading to the nearest neighbor match may be defined on such Z variables that have high explanatory power with respect to some of the Y and X variables. Nevertheless, it is highly unlikely that any distance function is optimal with respect to the entire set of Y variables. Therefore, the specific variables Y (or X) may be split into homogeneous groups; then for each group a single statistical match may be

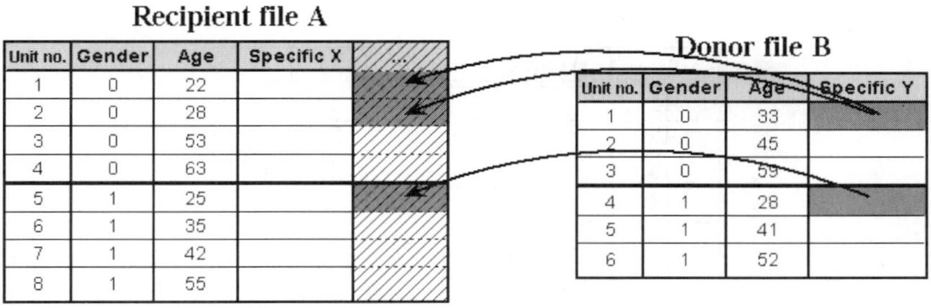

Figure 3.1. Matching classes by gender

performed. Sometimes it is convenient to use different common variables or to assign differing weights to them and search for the nearest neighbor donor according to each group. We call this matching in groups. In such cases, the additional variable Y (or X) of a unit of the statistically matched file can be imputed from more than one donor unit. This is illustrated in Figure 3.2.

3.3.1 Unconstrained Matching

If there are no restrictions on the number of times a certain donor unit may be imputed, then the statistical match is said to be unconstrained. This procedure has already been mentioned; we called it polygamy in Chapter 2. Unconstrained matching has the advantage of permitting the closest possible match for each record. However, the empirical mean and standard deviation of the imputed Y variable in the statistically matched file will usually not be identical with the corresponding statistic in the donor file.

We take up the artificial example as it was presented by Rodgers (1984) and also used for illustration purposes by Rubin (1986). Hence, we assume two small data files A and B of size $n_A = 8$ and $n_B = 6$, each being random samples from a population of $N = 24$ units as shown in Table 3.1 and already illustrated in Figures 3.1 and 3.2.

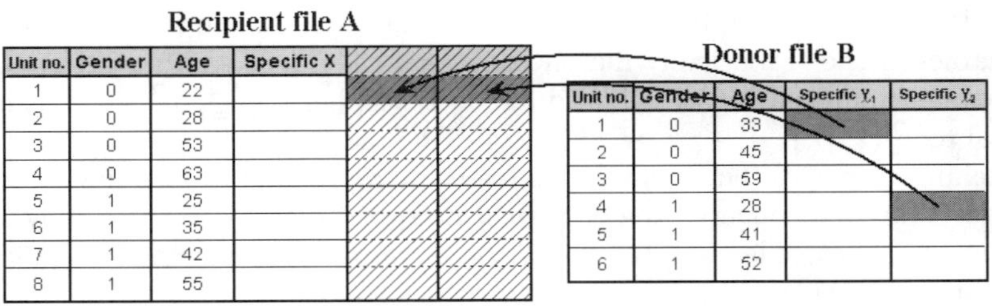

Figure 3.2. Matching in groups

File A

Unit i	Weight w_i^A	Z_1^A	Z_2^A	X
$A1$	3	1	42	x_1^A
$A2$	3	1	35	x_2^A
$A3$	3	0	63	x_3^A
$A4$	3	1	55	x_4^A
$A5$	3	0	28	x_5^A
$A6$	3	0	53	x_6^A
$A7$	3	0	22	x_7^A
$A8$	3	1	25	x_8^A

File B

Unit j	Weight w_j^B	Z_1^B	Z_2^B	Y
$B1$	4	0	33	y_1^B
$B2$	4	1	52	y_2^B
$B3$	4	1	28	y_3^B
$B4$	4	0	59	y_4^B
$B5$	4	1	41	y_5^B
$B6$	4	0	45	y_6^B

Table 3.1. Artificial example to illustrate unconstrained and constrained matching

Let variable Z_1 denote the gender of the respondents and Z_2 their age. As already discussed we require the match of gender to be mandatory forming a matching class: no match between male and female respondents is allowed. Within gender we look for a nearest neighbor unit by age where multiple use of donor units is allowed. The distance function d is defined to be the Euclidean distance (which equals the absolute difference for one single variable) in age between each pair; i.e., $d_{ij} = |z_{i2}^A - z_{j2}^B|$ is calculated between units $i = 1, 2, \ldots, 8$ of file A and units $j = 1, 2, \ldots, 6$ of file B. Those units with the minimum distance d are matched; for illustration see also Figure 3.1. Table 3.2 displays the final match.

It is obvious that after the match the empirical distributions of the Y variables in the statistically matched file and the donor sample differ. Calculating the mean of Y in the original file B yields $\frac{4}{24}\sum_{i=1}^{6} y_i^B$ which is not equal to $\frac{3}{24}(2y_1^B + y_2^B + y_3^B + 2y_4^B + 2y_5^B)$, the mean of Y observed after the matching in file A. The same holds for the empirical standard deviations. Notice that this concerns the empirical marginal distributions as they are observed in both files separately. In the example used by Rodgers (1984) the values of the Y variable are given with

$$y^B = (6.932 \; 5.524 \; 4.223 \; 6.147 \; 7.243 \; 3.230)'$$

Statistically matched file, recipient file A

Matched unit ij	Weight w_{ij}	Z_1^A	Z_2^A	Z_2^B	Distance d_{ij}	X	Y
$A1, B5$	3	1	42	41	1	x_1^A	y_5^B
$A2, B5$	3	1	35	41	6	x_2^A	y_5^B
$A3, B4$	3	0	63	59	4	x_3^A	y_4^B
$A4, B2$	3	1	55	52	3	x_4^A	y_2^B
$A5, B1$	3	0	28	33	5	x_5^A	y_1^B
$A6, B4$	3	0	53	59	6	x_6^A	y_4^B
$A7, B1$	3	0	22	33	11	x_7^A	y_1^B
$A8, B3$	3	1	25	28	3	x_8^A	y_3^B

Table 3.2. Statistically matched file due to unconstrained matching

having mean $\overline{y}^B = 5.55$ and standard deviation of $(1/5)\sum_{i=1}^{6}(y_i^B - \overline{y}^B)^2 = 1.57$. After the matching we find the following y-values transferred,

$$y^B = (7.243\ 7.243\ 6.147\ 5.524\ 6.932\ 6.147\ 6.932\ 4.223)'$$

now with mean $\overline{y}^B = 6.30$ and standard deviation $(1/7)\sum_{i=1}^{8}(y_i^B - \overline{y}^B)^2 = 1.04$. Notice that unit $B6$ is not matched to any unit of file A at all; thus, the value y_6^B is not represented in the statistically matched file. It has been viewed as a problem of the unconstrained match that the (empirical) marginal distributions of Y (or X, if file A is to be transferred to file B) in the statistically matched file may not be equal to their (empirical) marginal distributions in the original files; see Rodgers (1984) or Rubin (1986).

In Section 2.3.3 we focus on the theoretical marginal distributions which are the same in both files if the match is an unconstrained one and perfect in the common variables. There we show that under the assumption of

(1) using a donor unit without restrictions (i.e., unconstrained matching),

(2) identical Z variables when matched, and

(3) the separate files being random samples from the population of interest,

the statistically matched file can also be regarded as a random sample from the same population. Let us assume that the statistically matched file as well as the separate samples are drawn at random; then a simple two-sample t-test or F-test may be used to investigate whether two samples come from distributions with the same means or variances. If we apply these tests to our artificial example above we find no significant difference (p-value $= 0.3126$ for the F-test of equal variances; p-value $= 0.3017$ for the t-test of equal means (assuming equal variances)) between the distributions of Y as observed in file B and in the statistically matched file.

In Chapter 2 we discuss that conditional independence is established by this type of matching technique. To test whether X and Y given $Z = z$ are conditionally independent we may use the regression approach. Later in this work it is discussed in detail that testing the hypotheses of $\beta_{YX.Z} = 0$ from the multiple regression $Y = \beta_0 + \beta_{YZ_1.X} Z_1 + \beta_{YZ_2.X} Z_2 + \beta_{YX.Z} X + U$, $U \sim N_n(0, \sigma^2 I)$, is equal to testing the hypotheses of $\rho_{YX|Z} = 0$.[4] The values of the X variable are given with

$$x^A = (9.156\ 9.149\ 9.287\ 9.512\ 8.494\ 8.891\ 8.425\ 8.867)'$$

according to Rodgers (1984). If we apply the usual t-test for statistical significance of $\beta_{YX.Z}$ we find no conditional correlation between X and Y given $Z = z$ in the statistically matched file (p-value $= 0.8823$). Clearly this example is very simplified and the necessary assumptions for performing the standard test procedures are scarcely fulfilled; nevertheless some ideas may be clarified thereby.

Unconstrained matching is also called the "generalized distance method"; see Liu and Kovacevic (1998). In general, for each recipient record of file A a record from B is found such that the distance between them is minimal (within each matching class). The distance can be calculated using only the common variables Z or, in addition, some specific record weights w. Normalized z-values or their absolute values may be used. Different distance functions will lead to different matched files. The matching itself can be done searching for the nearest available match concerning this distance function. Let the values of p common variables of any unit i, $i = 1, 2, \ldots, n_A$ of file A be denoted by $z_i^A = (z_{i1}^A \quad z_{i2}^A \quad z_{ip}^A)'$ and of any unit j, $j = 1, 2, \ldots, n_B$ of file B $z_i^B = (z_{i1}^B \quad z_{i2}^B \quad z_{ip}^B)'$, respectively. Sometimes the City-Block metric is used taking the absolute differences according to

$$d_{ij}^{CB} = \sum_{k=1}^{p} |z_{ik}^A - z_{jk}^B|; \tag{3.1}$$

more often either the Euclidean distance given by

$$d_{ij}^E = \sqrt{(z_i^A - z_j^B)'(z_i^A - z_j^B)}, \tag{3.2}$$

or the Mahalanobis distance

$$d_{ij}^M = \sqrt{(z_i^A - z_j^B)' S_Z^{-1} (z_i^A - z_j^B)} \tag{3.3}$$

between the records based on the multivariate common variables Z is used (see Fahrmeir et al. (1996), pp. 449–450) for $i = 1, 2, \ldots, n_A$, $j = 1, 2, \ldots, n_B$, where S_Z is the estimated covariance matrix for the Z variables. For instance, S_Z can be calculated from a pooled or concatenated sample from files A and B; see Liu and Kovacevic (1998). Alternatively the mean value of a number of nearest neighbor units may be imputed or a donor unit chosen at random from a certain group of nearest neighbor

[4] For an extensive discussion see Section 5.2.3.

units. Therefore a tolerance t may be added to the observed minimum distance d_{ij} between two units i and j; all other donor units whose distances to i lie within that distance $d_{ij} + t$ may be used as potential candidates for imputation. Moreover, if there is more than one nearest neighbor unit, one of them may be selected at random.

However, some of the donor units may remain unmatched. The empirical marginal distribution of Y and the empirical conditional distribution of Y given $Z = z$ can be distorted; only the empirical distributions of Z and X as observed in file A are preserved. Nevertheless, discussing the validity of a matching technique in general, we found that the fourth level is obviously achieved because the marginal distributions of the donor file are sustained theoretically in the statistically matched file, although they may differ empirically. Usually it is to be expected that this difference is not greater than for any two samples drawn at random from the same population. Finally, classical unconstrained matching leads to conditional independence. How far the third and second levels are preserved depends on the explanatory power of the common variables to ensure conditional independence in the population of interest. In Section 3.3.5 we describe the British approach as a sophisticated version of unconstrained matching.

3.3.2 Constrained Matching

An alternative type of statistical matching is called constrained matching. The advantage of constrained matching (relative to unconstrained matching) is that the (empirical) multivariate distribution of the variables observed only in the donor file is precisely replicated in the statistically matched file. Analogous to Rodgers (1984) and Rubin (1986) we take up our simplified example to illustrate the basic principle of constrained matching before going into detail. According to Rodgers (1984) the conditions that must hold to preserve the margins by the (simple) constrained match can be written by means of the weights w_{ij}, with

$$\begin{aligned}\sum_{j=1}^{n_B} w_{ij} &= w_{i.}, \qquad \text{for } i = 1, 2, \ldots, n_A, \quad \text{and} \\ \sum_{i=1}^{n_A} w_{ij} &= w_{.j}, \qquad \text{for } j = 1, 2, \ldots, n_B.\end{aligned}$$

Again the distance function d_{ij} between units i, $i = 1, 2, \ldots, n_A$, in file A and units j, $j = 1, 2, \ldots, n_B$, in file B is calculated. The objective function of the problem is

$$\sum_{j=1}^{n_B} \sum_{i=1}^{n_A} d_{ij} w_{ij}, \qquad w_{ij} \geq 0, \quad i = 1, 2, \ldots, n_A, \quad j = 1, 2, \ldots, n_B, \qquad (3.4)$$

which has to be minimized according to the restrictions listed above. Liu and Kovacevic (1998) describe a linear programming approach as well as the so-called "generalized rank weight-split method" to solve this optimization task. The latter name comes from the fact that this method usually replicates some (or all) of the records from files A and B and, accordingly, splits their weights. Rubin (1986) gives a detailed description of how to "explode" the two files, to order them according to their z-values, and match them to get the final statistically matched file quickly; for illustration see Table 3.3.

File A				File B			
Unit i	Z_1^A	Z_2^A	X	Unit j	Z_1^B	Z_2^B	Y
$A7$	0	22	x_7^A	$B1$	0	33	y_1^B
$A7$	0	22	x_7^A	$B1$	0	33	y_1^B
$A7$	0	22	x_7^A	$B1$	0	33	y_1^B
$A5$	0	28	x_5^A	$B1$	0	33	y_1^B
$A5$	0	28	x_5^A	$B6$	0	45	y_6^B
$A5$	0	28	x_5^A	$B6$	0	45	y_6^B
$A6$	0	53	x_6^A	$B6$	0	45	y_6^B
$A6$	0	53	x_6^A	$B6$	0	45	y_6^B
$A6$	0	53	x_6^A	$B4$	0	59	y_4^B
$A3$	0	63	x_3^A	$B4$	0	59	y_4^B
$A3$	0	63	x_3^A	$B4$	0	59	y_4^B
$A3$	0	63	x_3^A	$B4$	0	59	y_4^B
$A8$	1	25	x_8^A	$B3$	1	28	y_3^B
$A8$	1	25	x_8^A	$B3$	1	28	y_3^B
$A8$	1	25	x_8^A	$B3$	1	28	y_3^B
$A2$	1	35	x_2^A	$B3$	1	28	y_3^B
$A2$	1	35	x_2^A	$B5$	1	41	y_5^B
$A2$	1	35	x_2^A	$B5$	1	41	y_5^B
$A1$	1	42	x_1^A	$B5$	1	41	y_5^B
$A1$	1	42	x_1^A	$B5$	1	41	y_5^B
$A1$	1	42	x_1^A	$B2$	1	52	y_2^B
$A4$	1	55	x_4^A	$B2$	1	52	y_2^B
$A4$	1	55	x_4^A	$B2$	1	52	y_2^B
$A4$	1	55	x_4^A	$B2$	1	52	y_2^B

Table 3.3. Exploding the data for constrained matching

Notice that this method uses only information about the Z ranks from files A and B and not their absolute values. Thus, the distance function is based on the ranks rather than on absolute values of Z.

In our example we multiply each unit by its weight to get a sample size of $N = 24$ in each sample. Then the units are ordered according to their common variables Z_1 and Z_2; this procedure is illustrated in Table 3.3.

Statistically matched file

Matched units ij	Weight w_{ij}	Z_1^A	Z_2^A	Z_2^B	Distance d_{ij}	X	Y
$A1, B2$	1	1	42	52	10	x_1^A	y_2^B
$A1, B5$	2	1	42	41	1	x_1^A	y_5^B
$A2, B3$	1	1	35	28	7	x_2^A	y_3^B
$A2, B5$	2	1	35	41	6	x_2^A	y_5^B
$A3, B4$	3	0	63	59	4	x_3^A	y_4^B
$A4, B2$	3	1	55	52	3	x_4^A	y_2^B
$A5, B1$	1	0	28	33	5	x_5^A	y_1^B
$A5, B6$	2	0	28	45	17	x_5^A	y_6^B
$A6, B4$	1	0	53	59	6	x_6^A	y_4^B
$A6, B6$	2	0	53	45	8	x_6^A	y_6^B
$A7, B1$	3	0	22	33	11	x_7^A	y_1^B
$A8, B3$	3	1	25	28	3	x_8^A	y_3^B

Table 3.4. Statistically matched file due to constrained matching

Calculating the statistically matched file is an easy task now. We match unit $A7$ three times with unit $B1$, unit $A5$ is matched once with unit $B1$ and twice with unit $B6$, and so on. Since many records have identical values we may store only those units with different values and summarize the results in a weighted file as shown in Table 3.4.[5] The new artificial statistically matched file is now of larger size than the separate samples A and B.

Now we see that the weighted average distance $\overline{d}$ between the matched cases is greater for this constrained match $\overline{d}^c = 155/24 = 6.46$ than for the unconstrained match $\overline{d}^u = 39/8 = 4.875$ as shown in Table 3.2. However, the mean as well as the standard deviation of the Y variable in the statistically matched file is equal to its value in file B because every unit of file B is imputed exactly four times into the exploded file A. Notice that the proper degrees of freedom from input file B rather than from the weighted statistically matched file should be used to calculate the estimated or empirical standard error of variable Y; see Rodgers (1984). In the weighted statistically matched file each value of variable Y is imputed twice; thus the estimated standard error in the constraint match has to be computed according to $(1/10)\sum_{i=1}^{12}(y_i^B - \overline{y}^B)^2 = 1.57$.

Finally, let us assume that the statistically matched file is drawn at random from the interesting population. Again the conditional correlation of X and Y given $Z = z$ in the statistically matched file is estimated to be zero by means of the regression approach (p-value $= 0.5637$). Although

[5]Notice that our results differ a bit from those published by Rodgers (1984) and Rubin (1986). Obviously the unit number of $A7$ was exchanged with $A5$ and vice versa; thus, their resulting matched file does not correspond with the procedure based on the original files. Therefore we present our calculation in detail.

constraints are used on the multiple use of donor units again conditional independence seems to be established in the matched file.

By means of constrained matching the empirical marginal distributions of X, Y, and Z are preserved in the statistically matched file. This matching technique is also sometimes called imputation "on rank"; see Liu and Kovacevic (1998). If the common variable Z is high-dimensional and the data set is large often this imputation on rank will be hard to implement. Then sorting the data according to some predefined order of univariate components of the common variables or according to a suitable categorical transformation of the components of variables is proposed. Moreover, the final match that is found depends on the order in which the common variables are used for building up the ranks. Also this procedure can lead to a rather large statistically matched file size. However, even matching under constraints establishes (more or less) conditional independence between the variables never jointly observed. In order to overcome the conditional independence assumption the use of auxiliary information is recommended.

3.3.3 Categorically Constrained Matching

In recent years more complicated versions of constrained matching have been applied to large data sets especially by Statistics Canada. Matching methods when an auxiliary data file is available were described and developed by Paass (1986), Singh et al. (1988, 1993), Liu and Kovacevic (1996, 1997, 1998), and Liu (1998). Singh et al. (1993) considered auxiliary information being available in the form of a categorical distribution. They proposed a modification of the loglinear imputation model as introduced by Singh et al. (1988). Liu and Kovacevic (1994, 1996, 1997) modified these methods to their categorically constrained matching procedure addressed to merge large survey files.

Principally the existence of an auxiliary data file C is assumed which contains records with x, y, and z-values or just x and y-values along with their survey weights if available. The quality of the information contained in this auxiliary data file is sometimes discussed (see Singh et al. (1993) and Liu and Kovacevic (1998)) but no practical guidelines are provided as to how to achieve such auxiliary data files.[6]

As already described in the context of unconstrained matching, the generalized distance method now may be applied using auxiliary information; see Kovacevic and Liu (1994). Therefore nearest neighbors are searched for each record of file A from the auxiliary file C first using a distance function

[6]It is sometimes pointed out that outdated or current files may be used; see Liu and Kovacevic (1998). We address this problem in Section 3.3.5 when the British approach is discussed. The German market research company GfK that extended the British solution additionally makes use of such auxiliary information derived from a small single source sample although it may be obviously biased.

of the variables Z and X which are now common to both files. Then the y-values from C are imputed in file A leading to a so-called "intermediate" statistically matched file. In the next step this intermediate file is matched with the donor file B using variables Z and Y as the common variables. Finally, the y-values from file B are imputed.

Using the weights from the auxiliary file leads to a similar adjustment when the constrained match (i.e., the rank weight-split method) is used. In a first step files A and C are exploded and sorted according to variables Z and X. Then the corresponding y-values are imputed and the intermediate file is created. In the second step the intermediate file is reordered according to Z and Y as is done with the exploded file B. Later the y-values are imputed according to the ranks and the final weights can be calculated from the multiply produced units.

Liu and Kovacevic (1996, 1997, 1998) extend these methods and build up a two-phase matching procedure that uses the auxiliary data file and deals with survey weights. In the first phase the matching is done with or without auxiliary variables and in the second phase a "minimum adjustment rematching algorithm" is applied to the statistically matched file. From the separate files A and B they calculate "categorical weights-sum tables" based on the (possibly categorized) variables Z, X and Z, Y, respectively. Depending on the auxiliary information available, an "auxiliary weights-sum table" is calculated using the noncommon variables (categorized) X and Y only or all (categorized) variables Z, X, Y. These tables are denoted and used as lookup tables in the matching or rematching process. The idea is to improve the categorical distribution of the matched file by the iterative adjustment of its margins to the margins of the separate files A and B and the auxiliary categorical weights table. Thus, the categorical associations from the separate files, i.e., between variables X and Z (file A) or Y and Z (file B), respectively, as well as from the categorical weights table concerning X and Y are preserved in the final statistically matched file. This may be done simply by weight adjustment or by an additional rematch through the application of the Minimum Shift-and-Share rematching algorithm where an iterative rematching of a small number of records is performed until the categorical constraints are fully satisfied; see Liu and Kovacevic (1998).

Basically the following steps are specified by Liu and Kovacevic (1997, 1998) to adjust the statistically matched file according to the categorical constraints given.[7]

(1) If they are not already of categorical scale, then the variables X, Y, Z involved in the matching process are transformed into categorical

[7]Notice that the separate files are already matched by some suitable procedure. The focus now is on improving the quality of the statistically matched file with regard to given categorical distributions.

variables $X^\star, Y^\star, Z^\star$. This is done by means of either some criteria for optimal partitioning or any auxiliary categorical information that may be available.

(2) Then the joint categorical distribution of $X^\star, Y^\star, Z^\star$ is estimated by raking the categorical distribution of the statistically matched file to available and adjusted marginal distributions derived from the separate files A and B as well as from the auxiliary file C. These estimated categorical distributions form the lookup tables.

(3) Finally, after the joint distribution of $X^\star, Y^\star, Z^\star$ is estimated, the individual weights of the records in this matched file are either adjusted or a partial rematching to satisfy the imposed constraints is performed and then the individual weights are adjusted if necessary.[8]

The reason why so much effort is made to ensure specific weights and categorical distributions is emphasized many times by Liu and Kovacevic (1994, 1996, 1997, 1998). They argue that surveys often contain individual record weights which are quite different; thus the creation of suitable record weights in the statistically matched file under the constraints of preserving the marginal or joint distributions given by an auxiliary data file is a difficult task especially with large data sets. Also stress is laid upon additional requirements that are imposed on statistical matching by the federal offices. For instance, the following requirements had to be satisfied for the creation of the Social Policy Simulation Database at Statistics Canada.

(1) The empirical conditional distribution of $Y|Z = z$ (as estimated by the donor file), or with the smallest possible distortion has to be preserved.

(2)) Then the use of all records from both files is required.

(3)) Finally, the size of the matched file should be modest; i.e., only a minimal inflation of the recipient file is allowed.

An extensive simulation study is presented by Liu and Kovacevic (1994, 1997, 1998) where constrained and unconstrained matching procedures with and without making use of an auxiliary data file are compared; also their categorical constraints rematching algorithm is used. Generally they find that the validity of any matching procedure (concerning our fourth-level validity) can be improved by additional categorical constraints. However, the quality of the lookup table (i.e., the quality of the auxiliary file) is important for their procedure. In the simulation study they have large data sets which are separated into numerous matching classes, but on

[8]The iterative proportional matching procedure which is implemented to synthesize and improve a statistically matched file according to such categorical lookup tables is described in detail by Liu (1998).

the other hand there are only a small number of specific variables. For example, the Y variable of file A is chosen to be univariate (e.g., the monthly gross rent or, alternatively, the owner's major payment) and the X variable is of bivariate kind (e.g., household investment income and household government transfer). Demographic information is used as common variables Z as are urbanization, residential tenure, household size and composition, and age and gender of the household maintainer. Furthermore the presence of a mortgage and a categorized total household income variable are chosen as common variable Z. To measure the preservation of the original relationship between X and Y the conditional correlation of X and Y when controlled for Z is computed. Actually they created four data files on which to base the simulation. In these original and complete data sets the conditional correlations between the two variables X and the variable Y given variables $Z = z$ are rather small already and a statistically significant conditional (partial) correlation is only found in four of eight cases. After the matches are performed again very small values of the conditional correlations are observed regardless of the applied matching procedure; see Liu and Kovacevic (1998). In their simulation the large number of matching classes implies only small differences between the records left within the classes. Thus, the methods under investigation could not differ much by default.

Finally, we find acknowledgment that all the methods based on a suitable nearest neighbor search tend to establish conditional independence whether auxiliary categorical information is used or not. The empirical marginal distributions of the separate files may be preserved better by use of additional information which concerns the fourth-level validity. However, until now little evidence has been given that the conditional independence assumption can be overcome by such procedures.

3.3.4 The Topological Concept

Contrary to the matching task as faced by Canada or the United States, in Europe usually separate surveys with rather large numbers of variables concerning any kind of media consumption and purchasing behavior are matched. The topological approach which is the basis of the German partnership model was designed by Wendt (1976, 1980, 1986). Although it is argued by Wendt and Wendt-Maeder (1983), pp. 10–20, that the concept is a topology-based approach, basically, it is a typology-based solution derived from cluster analysis procedures. Primarily this concept requires the use of as many common variables Z as are available. To start the fusion process, matching classes are formed according to variables that are defined as important a priori such as gender or county of residence. This segmentation is done because media consumption is said to be significantly different between males and females; furthermore, media supply has been quite different between the counties for a long time.

Within these matching classes the population of the donor sample B is divided by means of a typological analysis into homogeneous subgroups with regard to the common variables Z as well as the specific variables Y. To calculate these imputation types, cluster analysis algorithms are used which are based on a distance function being a mixture of the Euclidean distance similar to (3.2) for quantitative variables and the City-Block metric (absolute distance) (3.1) for any categorical variables that have to be coded with zero and one for each category. If a satisfying cluster solution or number of types are found, every individual is assigned to such a type. Now the types found in the donor sample are mapped into the recipient sample. Therefore the cluster means using only the common variables Z of each type are calculated and these cluster means are transferred into the recipient sample. To harmonize the separate samples every unit observed in the recipient sample is assigned to its nearest cluster mean based on the common variables.[9] According to the solution achieved in the recipient sample, new cluster means are calculated from the units assigned to the types in the recipient sample and transferred back into the donor sample. Again the assignment of the donor units is done according to the new cluster means in the donor sample. Transferring the cluster means and reorganizing the assignment twice is done to preserve the empirical joint distributions of the common and specific variables Z and Y in the final statistically matched file; see Koschnick (1995), p. 646.

Because in the first analysis of this iterative process the common and the specific variables of the donor sample are used to form the types, the cluster means usually differ considerably when they finally are transferred back. If some types are only filled with a few units then suitable types are aggregated. Also the assignment of the units to the types may not be unique; in other words, some individuals in the donor sample may also be very close to another type, or some units may be quite far away from their cluster mean. These units already assigned to a certain type are used to build up an additional reservoir to improve matching. Wendt (1976) called this reservoir of units that may also belong to another type its "aureole". The matching is done first within the types finally achieved: the types build up new matching classes. Based on the distance function for each unit of the recipient file the nearest neighbor donor according to the corresponding type is searched. Weights are also sometimes included in the distance function to form a common variable itself (survey weights) as well as to weight other common variables to make their match more likely. For each match a penalty weight is added to the donor unit to restrict its multiple use. Notice that many decisions during the whole matching procedure are

[9]Notice that every further calculation of distances and assignments uses the common variables Z only.

left to a personal choice rather than a mathematical rule; see Koschnick (1995), p. 647, or Wendt (1976).

After the first match is performed the units that are also members of an aureole are taken into account. Iteratively the overall distance between matched recipient and donor units is minimized by assigning donor units from the aureole rather than the type itself. If an average minimal distance of each type is found then all the specific values of Y are added to the recipient unit and the final statistically matched file is achieved.[10]

Due to the somewhat probabilistic assignment, especially of the members of the aureole, the topological concept seems to be rather similar to a procedure called "probabilistic cluster analysis" (see Bacher (2000)), which is equivalent to the generalized latent class approach (see Rässler and Fleischer (2000)). Since many rules and practical devices of the topological concept are left finally to the matching expert, the lack of mathematical foundation of this concept is evident and has always been emphasized by Wendt himself; see, e.g., Wendt (1979). However, to study the similarity between his concept and the latent class approach is beyond the scope of this work.[11] Basically the topological concept is a nearest neighbor matching procedure. The types are generated due to distance functions partly using even the specific variables. Thus, it is to be expected that the topological concept also establishes conditional independence between the variables never jointly observed given the common variables. All the tests that are performed and published (e.g., by Wendt (1976, 1979, 1980), Wiegand (1986), or Scheler and Wiegand (1987)) concern our fourth-level validity. The distributions as they are observed in the donor and the statistically matched file are not expected to differ more than they would if they were randomly drawn from the same population. Most of the tests applied and recommended by the AG.MA to prove the efficiency of this matching technique are based on tests concerning this difference; see also Section 2.5.5.

Since the use of many common variables is recommended the conditional correlation between the specific variables is expected to be very small in many cases. In this case this matching technique may yield more or fewer good results concerning the true association between the variables never

[10] The final rather complicated matching procedure has been implemented based on the programming language APL.

[11] The latent class approach applied to the statistical matching task has been developed by Kamakura and Wedel (1997) and is illustrated in a simulation study based on a real data set. Contrary to the traditional matching techniques, Kamakura and Wedel (1997) used the multiple imputation approach to create the statistically matched file multiply. The latent classes have the property of establishing conditional independence within the classes; thus, the basic idea is to estimate such classes and match the data within these latent classes to finally account for the conditional independence assumption. Although it is an interesting approach, its theory and practical application are very limited yet. This may be left to future research.

jointly observed. However, neither the use of an auxiliary file is provided nor is it possible to evaluate the validity of levels other than the fourth.

3.3.5 Multiple Ascription with Specific Variables

During the 1990s it became widely accepted that the validity of a statistically matched file depends on the power of the common variables to behave as good predictors of the specific information to be transferred; see Buck (1989), O'Brien (1991), Adamek (1994), and Roberts (1994). Especially in market research the limited extent of the demographic information to predict, for instance, the individual's propensity to watch television, became obvious. Consider, for example, the members of a television measurement panel. A woman working full time is likely to view less television than the average housewife. Nevertheless, her choice of programs is surely not well predicted from her demographic characteristics alone, for a detailed discussion see Roberts (1994).

Therefore, to efficiently link the viewing information from the British Broadcasters' Audience Research Board (BARB, the television measurement panel) with the purchasing data of the AGB Superpanel (a large market tracking panel which is conducted by the British company AGB), additional information about each Superpanel member's viewing habits is collected before matching the data. The claimed usual viewing behavior of the Superpanel members is used for matching with the usual metered viewing behavior of the BARB panel members. In an additional questionnaire emphasis is laid on comprehensibility to get the most accurate answers from the panel members. Hence, for this match measures of viewing behavior itself can be used as common variables and a great gain in validity is to be expected. As "specific common" variables viewing by day part and named programs are used to match Superpanel units with BARB donors. Of course, answers to the questions about day part and program viewing cannot be directly matched to the minute-by-minute viewing as metered exactly in the BARB data set. Therefore the BARB viewing information is aggregated and categorized within panel members with similar demographics. Finally, some 100 day part/program variables are achieved that are condensed to 30 principal component scores to serve as the viewing signature of each individual. Notice that the aim of the statistical matching here is to transfer the probability of viewing commercial breaks from the BARB donor sample to the Superpanel recipient sample. Therefore, in the next step the viewing signature is investigated by a series of multiple regression analyses to find out how well it may serve for predicting viewing commercial breaks. Weights for every element of the viewing signature variable are derived therefrom.

The following matching algorithm, called "multiple ascription" by Roberts (1994), is applied in detail.

(1) Based on the BARB sample for each day part and channel, the probability of viewing commercial breaks is regressed on the viewing signature and weights of importance for each component are derived therefrom.

(2) The weighted Euclidean distance according to (3.2) is calculated for every possible pair using only the viewing signature as the common variable Z.

(3) For each recipient the nearest three donor units are selected (i.e., a triple mean matching is applied). Finally, the mean of these three commercial viewing probabilities weighted with the inverse of each distance function is imputed to be the commercial viewing probability of a recipient for that day part and channel.

This matching algorithm is now performed for each day part and channel. Hence, the individual's total commercial viewing behavior as finally observed in the statistically matched file may come from many different donor units. Each imputed viewing information is calculated from three nearest neighbor units and matching in groups is performed; i.e., for each day part and channel new donor units are selected. Notice that no demographic information is taken as a common variable for the final match. Nevertheless, to achieve the artificial common variables (i.e., the viewing signature), demographic information is used in the first step. This new matching technique seems rather promising in improving the explanatory power of the common variables in the final matching process.

Extensive tests about the validity of this matching technique were reported by Roberts (1994). Here we find an interesting test addressing our first-level validity. A small recipient sample is taken from the large BARB sample and the remaining BARB units are used as donor units. From these remaining BARB units donor units are chosen, and the commercial viewing probability is calculated and imputed to these test-recipients according to the multiple ascription procedure as described above. Finally, the imputed commercial viewing probability is compared directly with the observed commercial viewing probability. This is similar to calculating the "counts" in our terminology (see Section 2.5) but only the preservation of the marginal distribution is tested. No conclusion can be drawn concerning the preservation of the joint distribution of the viewing and purchasing behavior.

However, conditional independence is still the main assumption for the multiple ascription approach. The use of the specific common variables suggests that their explanatory power is rather high. Thus, they may dissolve the joint association between the variables not jointly observed. This explicitly is the focus of the British multiple ascription approach; see Roberts (1994).

The biggest German market research company GfK has adopted this idea and extended it to their own matching procedure called MOVE.[12] Regularly the data from their consumer panel (ConsumerScan, based on about 12000 households that report their purchases weekly by means of an electronic diary) are matched with the data from a television research panel[13] where the viewing habits of about 5200 households are metered daily. Similar to the approach published by Roberts (1994) great importance is given to generating specific common variables. Therefore, all participants of the consumer panel are asked once a year to fill out a questionnaire reporting their preferred TV channel, viewing time, and genres. Likewise the members of the television measurement panel are questioned yearly about their main product usage. Similar to the procedure used by Roberts (1994) the corresponding specific variables from the more detailed panel surveys are aggregated and the distributions are adjusted to finally yield comparable specific common variables in each separate sample. Moreover, the GfK convinced about 500 households from the ConsumerScan also to participate in the television measurement panel. Clearly, this small single source panel is not representative for the population of interest. Nevertheless, these auxiliary data are used most efficiently to support the search for common variables that ensure the conditional independence assumption.

Furthermore, by means of factor analysis the specific variables Y are separated into homogeneous subgroups; for example, to impute the purchasing information from the ConsumerScan to the television measurement panel, seven to nine groups are defined such as food, hot drinks, nonalcoholic drinks, alcoholic drinks, etc. Using all the information available, such common variables are searched that ensure the conditional independence assumption between each consumer group variables and the television viewing information. Among other criteria these common variables Z get a rather high weight $g(Z)$ for use in the distance function. The final matching is done on the level of the household's maintaineer. Therefore, the distance between each recipient unit i and each possible donor unit j is calculated on the basis of p common variables $Z = (Z_1, Z_2, \ldots, Z_p)'$ according to

$$d_{ij} = \sqrt{\sum_{k=1}^{p} g(z_k)(z_{ki}^A - z_{kj}^B)^2 / \sum_{k=1}^{p} g(z_k)}\,, \tag{3.5}$$

for $i = 1, 2, \ldots, n_A$, $j = 1, 2, \ldots n_B$ where $g(z_k)$ is the individual weight of each variable Z_k, $k = 1, 2, \ldots, p$, within a purchasing group. The final matching in groups is done using some optimization routines trying to use each donor unit only once (monogamy with exceptions). Extensive tests

[12] The development of this matching procedure is supported by Bureau Wendt and us; the documentation was given to us with kind permission by the GfK.

[13] This television research panel is done on behalf of the AG.MA; the data are held private and confidential.

and experiments show rather encouraging results concerning our fourth as well as third and second level of validity.

However, the final aim of matching media and consuming data is to measure advertising effectiveness. Roberts (1994) reported detailed investigations which support that the matching technique can be used to evaluate the sales effectiveness of TV advertising on an existing large consumer market tracking panel. He finally states that "for brands with a reasonable high purchase frequency it offers a very cost effective solution to measuring both the effectiveness of TV advertising and the mechanics by which it works." Tests performed by the GfK support this hypothesis showing that due to the specific common variables advertising effectiveness may be measured in the matched sample although to a weakened degree; see Wildner (2000). A simulation study performed by Raetzel (2000), pp. 89–92, leaves the reader more ambivalent concerning measuring advertising effectiveness. Too many errors may occur due to measurement, sampling, or departures from the assumption of conditional independence. Obviously, advertising effectiveness can only be "transferred" by suitable specific common variables. If the explanatory power of the common and specific common variables is not very high, the quantification of pure advertising effectiveness seems to be very difficult if it is possible at all. However, again the focus is upon the assumption of conditional independence. It alone allows us to achieve a reliable statistically matched file.

3.4 Concluding Remarks

Rodgers (1984) lists some alternatives to the usual statistical matching procedures that would fully meet the objectives of statistical matching. Any analysis that requires only the univariate distributions of variables does not require statistical matching. They can be estimated from each file alone. Since all of the procedures establish conditional independence, statistical matching is also unnecessary whenever a covariance matrix is the objective of an analysis. Under the assumption of a multivariate normal distribution of X, Y, Z, for example, the assumption of conditional independence implies that $var(X, Y) = var(X, Z)'var(Z)^{-1}var(Z, Y)$ holds; see (2.17). Here $var(X, Z)$ can be estimated from file A, $var(Z, Y)$ from file B, and $var(Z)$ may be estimated either from one of the separate files A or B alone or from both, i.e., the pooled or concatenated files. If multivariate frequency tables concerning X and Y are involved in the analysis, they may be generated from the observed marginal and joint distributions by iterative proportional fitting. However, the flexibility a statistically matched file offers to the analyst makes it especially attractive. Also the separate files are usually high-dimensional making the calculation of alternatives more difficult or even impossible.

We have documented in this chapter that the matching techniques actually applied in practice are very sophisticated and are often the result of many years of experience. Much work is done before matching to prepare the separate files suitably; see Koschnick (1995), p. 645, or Liu and Kovacevic (1998). Often the samples have to be harmonized and variables and their distributions made comparable. Since all the techniques tend to establish conditional independence the preparation of the common and specific common variables is very important for maximizing their explanatory power. Extensive efforts have been spent over the years by institutes to do the matching in a way that minimizes the failure of the conditional independence assumption.

Rubin (1986) already states that "of particular importance, the optimal constrained approach suffers from the same major defect as the unconstrained approach, namely, it does not represent uncertainty about which value to impute due to either sampling variability or model uncertainty." All the matching techniques presented here and used in practice are searching for good common variables to ensure the conditional independence assumption. Procedures are well developed using as much information as is achievable. Nevertheless, none of the techniques we know of is able to overcome this assumption. Good matching results concerning our third and second level of validity are produced only if the common variables dissolve the association between the specific variables. Moreover, all the techniques suffer from the lack of methods to test and evaluate the quality of the common variables to serve as good predictors and improve the matching process. Until now we have not found any procedure to measure the predicting quality of the common variables if no really large auxiliary data file is available already containing the information needed. No measure is available to judge if we may "get close enough for any practical purposes."

To overcome both the assumption of conditional independence and the lack of measuring the validity, alternative approaches are presented in Chapter 4. Based on multiple imputation the range of the remaining possible unconditional correlation between the variables never jointly observed can be calculated enabling us to see whether the common variables may serve as good predictors by narrowing this range considerably. Even if the final matching is performed via traditional approaches, by means of the multiple imputation approach the explanatory power of the common variables may be evaluated beforehand. Also auxiliary information can be used most efficiently via multiple imputation to improve the matching process. Thus, we find it very promising to extend the ideas of Rubin (1986) and Manski (1995), trying to visualize the identification problem which is inherent in statistical matching. This is the focus of the following chapter.

4

Alternative Approaches to Statistical Matching

It is our experience that disagreements over identifying assumptions are often central to substantive controversies. Identifying assumptions that may seem credible to those making them are incredible to others. [...] We believe that the issues involved in many substantive controversies can be clarified, though certainly not resolved, by adopting data-analytic procedures that enable observers to distinguish between the range of conclusions permitted by the data and the specific conclusions that result from making specific identifying assumptions. We hope that analysts will adopt such approaches as complements to the standard procedure of making untestable and often controversial point-identifying assumptions.

J.L. Horowitz and C.F. Manski (2000a).

4.1 Introduction and Chapters Outline

Throughout this chapter we consider the matching problem as a problem of "file concatenation", a term coined by Rubin (1986); see Figure 4.1. First the two files A and B are concatenated and then the missing values of each part are multiply imputed to reflect uncertainty about the missing data and the unknown association of the variables never jointly observed. Thus our task is again to impute the missing data of X in file A and the missing data of Y in file B. U_{obs} in file A denotes the variables Z and Y and in file B, Z and X, respectively. Basically, this is a classical imputation problem.

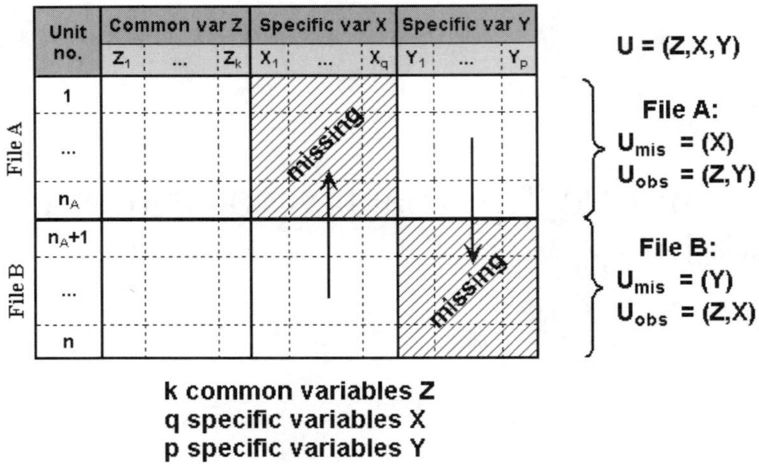

Figure 4.1. Statistical matching pictured as file concatenation

The practice of single imputation (i.e., filling in every missing datum with a more or less plausible value) has been used for decades now. For example, conditional mean imputation, and regression-based and hot deck procedures are widespread techniques for imputing missing information. All of the traditional matching techniques already presented herein refer to single imputation techniques, since for every unit of the recipient sample the missing information is imputed once from one or more units of the donor sample. Moreover these single imputation techniques are not based on random draws from some distributions. In recent years, however, imputations via random draws and thus multiple imputation (MI) techniques have emerged as a convenient and flexible tool for analyzing data with missing values.

The idea behind multiple imputation is that for each missing value we impute several values (e.g., m) instead of just one. These values may be generated according to some distributional assumptions concerning the missing data or, for example, in the statistical matching case by a simple random sample of m donor units from a pool of feasible donor units for each recipient unit.

These m imputations create m complete data sets; see Little and Rubin (1987). Each complete data set is analyzed using standard complete-data procedures just as if the imputed data were the real data. These results are later combined or pooled to produce estimates and confidence intervals that account for the missing data uncertainty according to the MI principle

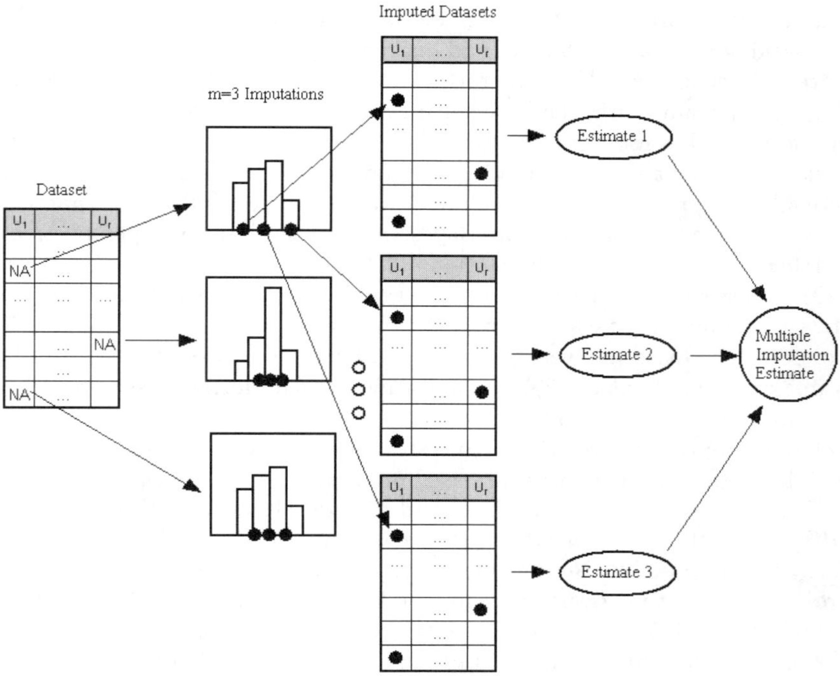

Figure 4.2. The concept of multiple imputation

defined by Rubin (1987). Figure 4.2 illustrates the concept of multiple imputation.[1]

The validity of the MI method depends on how the imputations were generated. Valid inference will scarcely be obtained if the imputations are created arbitrarily. An excellent introduction to the multiple imputation principle is given by Schafer (1999a). The concept of multiple imputation is also explained in detail by Brand (1999). In Section 4.3 we describe the multiple imputation principle focusing on statistical matching.

MI tries to bridge the gap between the Bayesian and the frequentist worlds. Inference from MI as proposed by Rubin (1987) is based on Bayesian statistical inference and on the other hand is validated according to some frequentist criteria. Rubin (1987), pp. 118–119, provides technical conditions under which a multiple imputation method leads to valid estimates from the frequentist point of view. Imputation techniques satisfying these conditions are said to be proper. Although these conditions are useful for

[1] Basically the figure is adopted from Brand (1999), p. 46, with some slight modifications. Notice that throughout the book missing data are denoted by "NA" (not available).

evaluating the properties of a given method, they provide little guidance for one seeking to create such a method in practice. Within the Bayesian framework, inference from MI is straightforward and much easier than within the frequentist framework. Thus, when introducing our multiple imputation method suitable for the statistical matching task, we create imputations under an explicit Bayesian model. As shown by Rubin (1987) this is theoretically and practically the most direct way to create imputations that lead to valid inference from both the Bayesian and the frequentist perspectives. Further discussions follow in Section 4.2.3.

Transferring MI procedures from the general missing data situation to the statistical matching task was proposed by Rubin (1986) and Rubin (1987), pp. 186–188. In recent years it has apparently only been readdressed by Kamakura and Wedel (1997). Imputations are performed assuming that the variables are conditionally independent within so-called latent classes. Yet their approach is restricted to restoring two-way contingency tables of variables never jointly observed based on a latent class model.[2]

The proposals of Raghunathani and Grizzle (1995) and Gelman et al. (1998) may also be connected with statistical matching although it is not their initial objective. Neither idea is designed to handle the identification problem inherent in statistical matching but the approach of Raghunathani and Grizzle (1995) seems to be an interesting path future work may follow. We discuss those approaches in the final chapter of the book.

The ability of the Bayesian framework to use prior information makes MI especially attractive for the statistical matching task, because it enables us to get estimates of the bounds of the unknown unconditional association. Hence, we may assume different conditional associations of the variables never jointly observed (given the common variables) to derive the range of their possible unconditional association. This idea was already introduced in Section 1.4 and dates back to Rubin (1987), p. 187.

Note that MI is not the only way of dealing with missing data in general nor is it necessarily the best method for any given problem; see, e.g., Little (1986). As discussed before, we focus on providing a so-called "public-use" data file generated from two (or even more) different sources. With the term public-use we want to stress the fact that the matched data are often passed forward to others; thus, data analyst and "imputer" will usually differ. Therefore our problem is a typical imputation task that may not be solved by other techniques such as weighting, calibration, or some EM algorithms.

This chapter is structured as follows. For the reader unfamiliar with the concepts of observed-data likelihood and observed-data posterior distribu-

[2] We have done some generalizations of their work (see Rässler and Fleischer (2000)), but there are still computational limitations for the general application on large data sets which is our main task here. Thus, we decided not to discuss this approach at present and leave it to future work.

tions the main assumptions are discussed first. An extensive and excellent discussion of these assumptions in the context of incompletely observed data in general may be found by the interested reader in Schafer (1997). We focus on the fact that neither the observed-data likelihood nor the traditional EM algorithm can handle the problem of statistical matching without an explicit assumption about the variables not jointly observed. Since Bayesian statistics provide some promising alternative approaches, such as the data augmentation algorithm which is very similar to the EM, we give a short introduction of Bayesian statistics applied to missing data. Examples and discussions hereinafter are always motivated by the statistical matching task; some essential derivations are given in the appendix. The multiple imputation principle is introduced then. Afterwards we discuss an imputation procedure with random residuals to perform MI, next this approach is generalized to a noniterative Bayesian multivariate linear regression model, then the data augmentation algorithm is described, and (last but not least) an approach based on regression switching is presented. Finally, a simulation study using a conventional multivariate normal model illuminates the advantages and disadvantages of the different techniques proposed.

4.2 Some Basic Notation

4.2.1 Ignorability of the Missing Data Mechanism

To simplify the presentation we assume no additional missing data among the variables in file A or B, respectively. The sample design underlying both samples is assumed to be a simple random sample. As a consequence of these assumptions, files A and B can be considered as repeated independent samples from the same population.

Following the terminology introduced by Rubin (1987) and Little and Rubin (1987) the missing-data mechanism can be classified according to the probability of response. The missing data are said to be as follows.

- MCAR – missing completely at random. In this case, the unobserved values form a random subsample of the sampled values. If, for instance, the probability that income is recorded is the same for all individuals, regardless of, e.g., their age or income itself, then the data are said to be MCAR.

- MAR – missing at random. In this case, the unobserved values form a random subsample of the sampled values within subclasses defined by the observed values. For example, if the probability that income is recorded varies according to the age of the respondent but does not vary according to the income of the respondent within an age group, then the data are MAR.

- MNAR – missing not at random. If the data are neither MCAR nor MAR, the mechanism is nonignorable. If the probability that income is recorded varies according to income itself, then the data are MNAR.

In the first two cases, when the data are MCAR or MAR, these missing-data mechanisms are ignorable for likelihood-based inference. In particular, a missing data mechanism is said to be ignorable if (1) the missing data generation mechanism is not relevant for inference on parameters describing the complete data, and (2) the mechanism generating the missing data depends only on the observed data and not on the missing data. Clearly, the nonignorable case is the hardest to deal with analytically. Concerning statistical matching we agree with Kamakura and Wedel (1997) as well as Gelman et al. (1998) to regard the missing data, at least, as missing at random. It seems to be a reasonable assumption here, because all of the missingness is due to unasked questions. The missingness is induced by the study design. Thus, this particular missing data mechanism is deterministic and may be regarded as ignorable.

4.2.2 Observed-Data Likelihood

Definition

Let $\mathcal{U}$ be any population of interest, finite or not, and $u_i = (u_{i1}, u_{i2}, \ldots, u_{ir})$ denote the value of a random variable $U = (U_1, U_2, \ldots, U_r)$ for each unit $i \in \mathcal{U}$. Without loss of generality let $f_U(u_i; \theta)$ be the probability of drawing a certain unit i, $i \in \mathcal{U}$ with observations $u_i = (u_{i1}, u_{i2}, \ldots, u_{ir})$ depending on the parameter θ which may be regarded as a scalar or vector. In the case of continuous random variables U, f_U may be taken as the density function instead of the probability function. To be more general, f_U may also describe a finite mixture of densities. Finally, let a random sample of n independently observed units from $\mathcal{U}$ be given with probability or, more generally, with density function $\prod_{i=1}^{n} f_U(u_i; \theta)$.

Now denote the observed part of the random variable U by U_{obs}, and the missing part by U_{mis}, so that $U = (U_{obs}, U_{mis})$. The joint distribution of U_{obs} and U_{mis} is given by $f_U(u_i; \theta) = f_{U_{obs}, U_{mis}}(u_{obs,i}, u_{mis,i}; \theta)$ for each unit $i \in \mathcal{U}$.

Furthermore, let R be an indicator variable being zero or one depending on whether the corresponding element of U is missing or observed; i.e.,

$$R_{ij} = \begin{cases} 1 & \text{, if variable } U_j \text{ is observed on the unit } i, \\ 0 & \text{, else,} \end{cases}$$

for all units $i \in \mathcal{U}$ and variables $U_j, j = 1, 2, \ldots, r$. Generally a probability model for R with $f_{R|U}(r|u; \xi, \theta)$ is assumed, which depends on U and θ as well as some unknown nuisance parameter ξ. Hence, the joint distribution of the response indicator R and the interesting variables U is given by $f_{R,U}(r, u; \xi, \theta)$. The density or probability function describing the observed data of any unit $i \in \mathcal{U}$ and, thus, their likelihood may actually be written

$$L(\theta, \xi; u_{obs}, r) = f_{R,U_{obs}}(r, u_{obs}; \xi, \theta) = \int f_{R|U}(r|u; \xi, \theta) f_U(u; \theta) du_{mis}. \tag{4.1}$$

For simplicity we want the integral to be understood as the sum for discrete distributions. To ease reading, we usually refer to f_U as the density function of U hereinafter.

Now the assumptions concerning the missing-data mechanisms can be explained more technically.

- First of all it is assumed that θ and ξ are "distinct"; i.e., knowing θ will provide no information about ξ and vice versa (see Schafer (1997), p. 11). This allows us to proceed further and assume that the conditional distribution of R given a value $U = u$, i.e., $R|U = u$ or $R|u$, does not depend on θ and $f_{R|U}(r|u; \xi, \theta) = f_{R|U}(r|u; \xi)$ holds.
- Under the MCAR mechanism the response indicator R and the interesting variables U are assumed to be independent with $f_{R|U}(r|u; \xi) = f_R(r; \xi)$.
- Under the MAR mechanism the conditional distribution of $R|u$ does not depend on the missing data U_{mis} and is given by $f_{R|U}(r|u; \xi) = f_{R|U_{obs}}(r|u_{obs}; \xi)$.

Thus we have seen, if the parameters ξ and θ are distinct and the missing-data mechanism is at least MAR, then the conditional distribution of $R|u$ is given by $f_{R|U}(r|u; \xi, \theta) = f_{R|U_{obs}}(r|u_{obs}; \xi)$. The conditional distribution of $R|u$ is independent of U_{mis} and θ; the missingness mechanism is said to be ignorable. The likelihood (4.1) of the observed data of any unit $i \in \mathcal{U}$ under MAR can now be factorized into

$$\begin{aligned} L(\theta, \xi; u_{obs}, r) &= f_{R,U_{obs}}(r, u_{obs}; \xi, \theta) = f_{R|U_{obs}}(r|u_{obs}; \xi) f_{U_{obs}}(u_{obs}; \theta) \\ &= f_{R|U_{obs}}(r|u_{obs}; \xi) \underbrace{\int f_U(u; \theta) du_{mis}}_{= L(\theta; u_{obs,i})} . \end{aligned} \tag{4.2}$$

According to Little and Rubin (1987), pp. 88–92, and illustrated by (4.2) under ignorable missingness it is not necessary to consider a model for R if likelihood-based inference about θ is intended. Given n observations independently drawn from the underlying population, the likelihood ignoring the missing-data mechanism is

$$\begin{aligned} L(\theta; u_{obs}) &= \prod_{i=1}^{n} L(\theta; u_{obs,i}) = \prod_{i=1}^{n} f_{U_{obs}}(u_{obs,i}; \theta) \\ &= \prod_{i=1}^{n} \int f_U(u_i; \theta) du_{mis,i} . \end{aligned} \tag{4.3}$$

In our matching or file concatenation case the observed-data likelihood is obviously given by

$$\begin{aligned} L(\theta; u_{obs}) &= \prod_{i=1}^{n_A} f_{Z,Y}(z_i, y_i; \theta) \prod_{i=n_A+1}^{n} f_{Z,X}(z_i, x_i; \theta) \\ &= \prod_{i=1}^{n_A} \int f_{Z,X,Y}(z_i, x_i, y_i; \theta) dx_i \\ &\quad \prod_{i=n_A+1}^{n} \int f_{Z,X,Y}(z_i, x_i, y_i; \theta) dy_i. \end{aligned} \tag{4.4}$$

Actually this is the observed-data likelihood ignoring the missing-data mechanism. Notice that we focus on the matching case assuming an ignorable missing data mechanism throughout this work; thus, for brevity we refer to (4.3) and (4.4) simply as the observed-data likelihood.

Hence we have seen that all relevant statistical information about the parameters incorporated by θ should be contained in the observed-data likelihood $L(\theta; u_{obs})$, if the complete-data model and the ignorability assumptions are correct. Although our assumptions may be valid, we recognize for the matching case by (4.4) that no inference from the observed-data likelihood is derived about the association of the variables (X, Y) never jointly observed. However, the well-known EM algorithm is designed to solve difficult incomplete-data problems by repeatedly solving tractable complete-data problems. Although this approach is intuitively appealing, the EM does not work in the matching case. We lay stress upon this result in the following sections.

EM Algorithm

The Expectation-Maximization (EM) algorithm is a broadly applicable approach to the iterative computation of maximum likelihood estimates. It is useful in a variety of incomplete data problems, where, for example, the Newton–Raphson method may turn out to be more complicated. Especially in the context of missing data, when the observed-data likelihood (4.3) is difficult to derive, the EM may be a helpful alternative at hand.

On each iteration of the EM algorithm there are two steps, called the expectation or E-step and the maximization or M-step. Roughly speaking, the basic idea of the EM algorithm is first (E-step) to fill in the missing data U_{mis} by their conditional expectation given the observed data and an initial estimate of the parameter θ to achieve a completed likelihood function, and second (M-step) to recalculate the maximum likelihood (ML) estimate of θ given the observed values u_{obs} and the filled-in values of $U_{mis} = u_{mis}$. Then the E-step and M-step are iterated until convergence of the estimates is achieved.

More precisely, it is the log likelihood $\ln L(\theta; u)$ of the complete-data problem that is manipulated in the E-step. As it is based partly on unobserved data, it is replaced by its conditional expectation

$$E(\ln L(\theta; U)|u_{obs}; \theta^{(t)})$$

given the observed data u_{obs} and a current fit $\theta^{(t)}$ for the unknown parameters. Thus the E-step consists of calculating this conditional expectation $E(\ln L(\theta; U)|u_{obs}; \theta^{(t)})$. The simpler M-step computation can now be applied to this completed data and a new actual value $\theta^{(t+1)}$ for the ML estimate is computed therefrom. Now let $\theta^{(t+1)}$ be the value of θ that maximizes $E(\ln L(\theta; U)|u_{obs}; \theta^{(t)})$. Dempster et al. (1977) have shown that $\theta^{(t+1)}$ then also maximizes the observed-data likelihood $L(\theta; u_{obs})$ in the sense that the observed-data likelihood of $\theta^{(t+1)}$ is at least as high as that of $\theta^{(t)}$; i.e., $L(\theta^{(t+1)}; u_{obs}) \geq L(\theta^{(t)}; u_{obs})$ holds.

Starting from some suitable initial parameter values $\theta^{(0)}$, the E- and the M-steps are repeated until convergence, for instance, until $|\theta^{(t+1)} - \theta^{(t)}| \leq \epsilon$ holds for some fixed $\epsilon > 0$. Not all the problems are well-behaved, however, and sometimes the EM does not converge to a unique global maximum as can be shown in the matching case. For a detailed description of the EM algorithm and its properties the interested reader is referred to McLachlan and Krishnan (1997), Schafer (1997), Little and Rubin (1987), and the fundamental paper of Dempster et al. (1977). We therefore do not go into further technical details here.

Illustration with Normally Distributed Data

This example has two objectives. The first is to demonstrate the shortcomings of the EM when applied to the matching situation. Then we prepare to highlight the similarities between the EM and the data augmentation algorithms in the later section, because data augmentation in general can be viewed as the Bayesian stochastic counterpart of the EM. Similar examples are given by McLachlan and Krishnan (1997), Schafer (1997), and Little and Rubin (1987).

Now, let $(Z, X, Y) \sim N_3(\mu, \Sigma)$, with

$$\mu = \begin{pmatrix} \mu_Z \\ \mu_X \\ \mu_Y \end{pmatrix} \text{ and } \Sigma = \begin{pmatrix} \sigma_{ZZ} & \sigma_{ZX} & \sigma_{ZY} \\ \sigma_{ZX} & \sigma_{XX} & \sigma_{XY} \\ \sigma_{ZY} & \sigma_{XY} & \sigma_{YY} \end{pmatrix};$$

hence the joint distribution of $U = (Z, X, Y)$ with $u = (z, x, y)'$ is given by

$$f_U(u; \mu, \Sigma) = (2\pi)^{-3/2}|\Sigma|^{-1/2} \exp\left\{-\frac{1}{2}(u - \mu)'\Sigma^{-1}(u - \mu)\right\}.$$

Notice that we usually denote the covariance matrix of some random variables with Σ and the covariance between two random variables with σ throughout this work.

If we consider the file concatenation problem with

$$f_{U_{obs}} = f_{Z,Y}(z, y; \mu_Z, \mu_Y, \Sigma_{ZY})$$

for the units observed in file A and with

$$f_{U_{obs}} = f_{Z,X}(z, x; \mu_Z, \mu_X, \Sigma_{ZX})$$

for the units observed in file B, we calculate the observed-data loglikelihood (assuming MAR and ignorable missingness) according to

$$\begin{aligned}
\ln L(\mu, \Sigma; u_{obs}) &= \sum_{i=1}^{n} \ln f_{U_{obs}} \\
&= \sum_{i=1}^{n_A} \ln f_{Z,Y}(z_i, y_i; \mu_Z, \mu_Y, \Sigma_{ZY}) + \sum_{i=n_A+1}^{n} \ln f_{Z,X}(z_i, x_i; \mu_Z, \mu_X, \Sigma_{ZX}) \\
&= -\frac{3}{2} n \ln(2\pi) - \frac{n_A}{2} \ln |\Sigma_{ZY}| \\
&\quad -\frac{1}{2} \sum_{i=1}^{n_A} (z_i - \mu_Z \quad y_i - \mu_Y) \Sigma_{ZY}^{-1} \begin{pmatrix} z_i - \mu_Z \\ y_i - \mu_Y \end{pmatrix} \\
&\quad -\frac{n_B}{2} \ln |\Sigma_{ZX}| - \frac{1}{2} \sum_{i=n_A+1}^{n} (z_i - \mu_Z \quad x_i - \mu_X) \Sigma_{ZX}^{-1} \begin{pmatrix} z_i - \mu_Z \\ y_i - \mu_Y \end{pmatrix},
\end{aligned}$$

$$\text{with} \quad \Sigma_{ZX} = \begin{pmatrix} \sigma_{ZZ} & \sigma_{ZX} \\ \sigma_{ZX} & \sigma_{XX} \end{pmatrix} \quad \text{and} \quad \Sigma_{ZY} = \begin{pmatrix} \sigma_{ZZ} & \sigma_{ZY} \\ \sigma_{ZY} & \sigma_{YY} \end{pmatrix}.$$

Of course, for this small example the ML estimates based on the observed data can easily be derived from this observed-data loglikelihood. For demonstration purposes we discuss the complete case analysis which results in well-known statistics based on the complete-data likelihood:

$$\begin{aligned}
\ln L(\theta; u) &= \sum_{i=1}^{n} \ln f_U(u_i; \mu, \Sigma) \\
&= -\frac{3}{2} n \ln(2\pi) - \frac{n}{2} \ln |\Sigma| - \frac{1}{2} \sum_{i=1}^{n} (u_i - \mu)' \Sigma^{-1} (u_i - \mu),
\end{aligned}$$

with $u_i = (z_i \quad x_i \quad y_i)'$ and $\mu = (\mu_Z \quad \mu_X \quad \mu_Y)'$, $i = 1, 2, \ldots, n$.

The complete-data sufficient statistics T derived therefrom[3] are given by

$$T_j = \sum_{i=1}^{n} U_{ji} \text{ and } T_{jl} = \sum_{i=1}^{n} U_{ji}U_{li}, \text{ with } \begin{cases} j,l=1 & \text{denoting } Z_i, \\ j,l=2 & \text{denoting } X_i, \\ j,l=3 & \text{denoting } Y_i. \end{cases} \tag{4.5}$$

The usual ML estimates based on the complete data are

$$\begin{aligned} \widehat{\mu}_j &= T_j/n\,, \qquad j=1,2,3\,, \\ \widehat{\sigma}_{jl} &= (T_{lj} - n^{-1}T_jT_l)/n\,, \qquad j,l=1,2,3\,. \end{aligned} \tag{4.6}$$

To achieve the conditional expectation $E(\ln L(\mu,\Sigma;U)|u_{obs})$ of the complete-data likelihood, we calculate the expectations $E(T|u_{obs})$ of the sufficient statistics given the observed data. More precisely we have to derive the expectations $E(X_i|Z=z_i, Y=y_i)$ and $E(X_i^2|Z=z_i, Y=y_i)$, $i=1,2,\ldots,n_A$ for file A as well as $E(Y_i|Z=z_i, X=x_i)$ and $E(Y_i^2|Z=z_i, X=x_i)$, $i=n_A+1,\ldots,n$ for file B.

From the well-known properties of the multivariate normal distribution, the conditional distribution of the missing variable X given $Z=z$ and $Y=y$ (and likewise Y given $Z=z$ and $X=x$) is univariate normal assuming independent observations in each row with

$$\begin{aligned} &(\text{file } A) \quad X_i|z_i,y_i \sim N_1\left(\mu_{X_i|Z_iY_i}, \sigma^2_{X|ZY}\right), \; i=1,2,\ldots,n_A\,, \\ &(\text{file } B) \quad Y_i|z_i,x_i \sim N_1\left(\mu_{Y_i|Z_iX_i}, \sigma^2_{Y|ZX}\right), \; i=n_A+1,\ldots,n. \end{aligned} \tag{4.7}$$

These conditional distributions $f_{U_{mis}|U_{obs}}(u_{mis}|u_{obs};\mu,\Sigma)$ of the missing data given the observed data are the so-called (conditional) predictive distributions of the missing data.

For file A the parameters of the predictive distributions are calculated with

$$\begin{aligned} \mu_{X_i|Z_iY_i} &= \mu_X + (z_i - \mu_Z\,, \quad y_i - \mu_Y)\Sigma_{11}^{-1}\Sigma_{12}\,, \quad i=1,2,\ldots,n_A, \\ \sigma^2_{X|ZY} &= \sigma^2_X - \Sigma_{12}'\Sigma_{11}^{-1}\Sigma_{12} \text{ where} \\ \Sigma_{11} &= \begin{pmatrix} \sigma_{ZZ} & \sigma_{ZY} \\ \sigma_{ZY} & \sigma_{YY} \end{pmatrix}, \quad \Sigma_{12} = \begin{pmatrix} \sigma_{ZX} \\ \sigma_{XY} \end{pmatrix}. \end{aligned} \tag{4.8}$$

For file B the corresponding values are computed according to

$$\mu_{Y_i|Z_iX_i} = \mu_Y + (z_i - \mu_Z\,, \quad x_i - \mu_X)\Sigma_{11}^{-1}\Sigma_{12}\,, \; i=n_A+1,\ldots,n,$$

[3]It can be shown that the complete-data loglikelihood belongs to the exponential family with sufficient statistics T. Loosely speaking, a sufficient statistic is a function such as $\sum U_i$ of a sample $U_1, U_2, \ldots, U_n$ that incorporates just as much information about the parameter of interest, e.g., μ_U, as the sample itself. For further details see Schafer (1997), pp. 39–42, McLachlan and Krishnan (1997), pp. 45–49, or Mood et al. (1974), pp. 299–314.

$$\sigma^2_{Y|ZX} = \sigma^2_Y - \Sigma'_{12}\Sigma^{-1}_{11}\Sigma_{12}\,, \text{ where} \tag{4.9}$$
$$\Sigma_{11} = \begin{pmatrix} \sigma_{ZZ} & \sigma_{ZX} \\ \sigma_{ZX} & \sigma_{XX} \end{pmatrix}, \quad \Sigma_{12} = \begin{pmatrix} \sigma_{ZY} \\ \sigma_{XY} \end{pmatrix}.$$

Notice that with a little abuse of notation we sometimes do not distinguish between random variables and their values, especially when matrices are used. We usually denote matrices, e.g., Z, with capital letters whether random variables or not. When, on the other hand, we condition on variables, e.g., X given $Z = z$, we denote them with small letters, e.g., $X|z$. Moreover, indices are written in capital letters throughout the book although they may correspond to values of random variables; i.e., instead of the correct formulation $\mu_{Y_i|Z_i=z_i,X_i=x_i}$ we use $\mu_{Y_i|Z_iX_i}$ or even shorter $\mu_{Y|ZX}$.

As long as there are no data completely observed for X and Y the conditional association of X and Y given $Z = z$ cannot be estimated from the observed data. Usually the common variables Z determine neither X nor Y exactly, and, hence, the unconditional association of X and Y will be bounded but not identifiable. In such cases the EM algorithm may quickly converge, but unfortunately the EM solutions are not unique. Different starting values will converge to different estimates. The likelihood is said to have a ridge, and the EM algorithm stops once the ridge is reached; see Schafer (1997), p. 53.

To illustrate the application of the EM algorithm to this type of problem, we now apply it to the data set in Table 4.1, which was used by McLachlan and Krishnan (1997), p. 49, in a similar form. It is enlarged here to demonstrate our problem. There are $n = 10$ trivariate observations with X and Y never jointly observed; the missing values are marked by NA.

Applying the EM algorithm to this data set, we record different values for σ_{XY} depending on its starting values. Table 4.2 displays the convergence achieved after 231 iterations of the EM with a value of $\sigma^{(\infty)}_{XY} = 4.45$ and $\rho^{(\infty)}_{XY} = 0.2062$.

Other starting values lead to convergence after 589 iterations (see Table 4.3), with $\sigma^{(t)}_{XY}$ converging to $\sigma^{(\infty)}_{XY} = 6.33$ and $\rho^{(\infty)}_{XY} = 0.2933$.

Using, for example, the identity matrix as the starting value for Σ, $\sigma^{(t)}_{XY}$ converges to a value of $\sigma^{(\infty)}_{XY} = -1.16$ and $\rho^{(\infty)}_{XY} = -0.0537$.

i	1	2	3	4	5	6	7	8	9	10
$Z = z_i$	8	11	16	18	6	4	20	25	9	13
$X = x_i$	NA	NA	NA	NA	NA	NA	1	5	8	12
$Y = y_i$	10	14	16	15	20	4	NA	NA	NA	NA

Table 4.1. Arbitrary trivariate data set

t	$\mu_Z^{(t)}$	$\mu_X^{(t)}$	$\mu_Y^{(t)}$	$\sigma_{ZZ}^{(t)}$	$\sigma_{ZX}^{(t)}$	$\sigma_{ZY}^{(t)}$	$\sigma_{XX}^{(t)}$	$\sigma_{XY}^{(t)}$	$\sigma_{YY}^{(t)}$
1	13.00	12.00	14.00	1.00	0.50	0.20	1.00	0.50	1.00
2	13.00	12.08	11.47	40.20	6.32	-16.17	18.91	13.24	27.38
3	13.00	11.92	10.87	40.20	4.29	-19.73	22.62	17.35	33.49
4	13.00	12.00	10.66	40.20	4.92	-19.89	22.92	16.52	32.32
5	13.00	12.11	10.49	40.20	5.65	-19.88	22.89	15.33	30.76
6	13.00	12.21	10.33	40.20	6.36	-19.85	22.86	14.20	29.33
7	13.00	12.32	10.18	40.20	7.04	-19.80	22.88	13.16	28.04
8	13.00	12.42	10.04	40.20	7.70	-19.74	22.94	12.22	26.88
9	13.00	12.53	9.91	40.20	8.34	-19.66	23.04	11.36	25.84
10	13.00	12.62	9.78	40.20	8.94	-19.58	23.17	10.58	24.92
...	...	...	...	...	...	...	...	...	...
∞	13.00	14.24	8.03	40.20	17.19	-16.45	28.08	4.45	16.59

Table 4.2. Results of the EM algorithm with starting covariance 0.5

The higher we set the starting value for σ_{XY} the longer convergence takes. With $\sigma_{XY}^{(1)} = 0.9$ instead of 0.5 in tTable 4.2 we finally reach convergence after 7258 iterations with $\sigma_{XY}^{(\infty)} = 7.1994$ and $\rho_{XY}^{(\infty)} = 0.3336$. If we use the calculated values from Table 4.2 but not $\widehat{\sigma}^2_{X,Y}$ as estimates for the covariance or correlation matrix, we get

$$\widehat{cov}(Z,X,Y) = \begin{pmatrix} 40.20 & 17.19 & -16.45 \\ 17.19 & 28.08 & \widehat{cov}(X,Y) \\ -16.45 & \widehat{cov}(X,Y) & 16.59 \end{pmatrix} \quad \text{with}$$

$$\det(\widehat{cov}(Z,X,Y)) = -40.20\widehat{cov}(X,Y)^2 - 2 \cdot 282.78\widehat{cov}(X,Y) + 6226.26.$$

Calculating the possible range of the estimate $\widehat{cov}(X,Y)$ of the unconditional association by means of the roots of $\det(\widehat{cov}(Z,X,Y))$ gives us $\widehat{cov}(X,Y) \in [-7.2613; 21.3297]$ or $\widehat{\rho}_{XY} \in [-0.3364; 0.9882]$, respectively.

t	$\mu_Z^{(t)}$	$\mu_X^{(t)}$	$\mu_Y^{(t)}$	$\sigma_{ZZ}^{(t)}$	$\sigma_{ZX}^{(t)}$	$\sigma_{ZY}^{(t)}$	$\sigma_{XX}^{(t)}$	$\sigma_{XY}^{(t)}$	$\sigma_{YY}^{(t)}$
1	10.00	10.00	10.00	1.00	0.60	0.50	1.00	0.80	1.00
2	13.00	11.69	10.09	40.20	1.01	-9.55	20.44	20.13	24.99
3	13.00	11.64	9.94	40.20	-1.03	-12.29	23.05	23.08	28.61
4	13.00	11.68	9.89	40.20	-0.76	-12.50	23.00	22.64	28.03
5	13.00	11.72	9.84	40.20	-0.46	-12.66	22.88	22.14	27.38
6	13.00	11.76	9.79	40.20	-0.16	-12.82	22.77	21.64	26.76
7	13.00	11.80	9.75	40.20	0.15	-12.97	22.68	21.16	26.17
8	13.00	11.84	9.71	40.20	0.45	-13.12	22.59	20.69	25.61
9	13.00	11.88	9.67	40.20	0.75	-13.26	22.51	20.23	25.08
10	13.00	11.92	9.62	40.20	1.05	-13.40	22.44	19.78	24.58
...	...	...	...	...	...	...	...	...	...
∞	13.00	14.24	8.03	40.20	17.19	-16.45	28.08	6.33	16.59

Table 4.3. Results of the EM algorithm with starting covariance 0.8

The matrix $\widehat{cov}(Z, X, Y)$ yields a feasible covariance matrix only within this range of $\widehat{cov}(X, Y)$ if all other variances and covariances are held fixed. Apparently we will not derive the bounds of the unconditional association of the variables never jointly observed by means of the EM. Finally, we are not able to obtain something useful via the EM as long as we do not have access to a small but completely observed data source stabilizing the estimation of the parameters and leading to unique solutions. Moreover our interest is less in the unbiased estimation of the model parameters than in the proper imputation of the missing values to achieve a public-use file, which is obviously not possible by means of the EM.

As a matter of fact, we thus introduce some prior information using Bayesian statistics.

4.2.3 Bayesian Inference

The Bayesian paradigm is based on specifying a probability model for the observed data U with joint density $f_{U|\Theta}(u|\theta)$ given a vector of unknown parameters $\Theta = \theta$ which is identical to the likelihood function $L(\theta; u)$ understood as a function of θ. Then we assume that Θ is random[4] and has a prior distribution with density or probability function f_Θ. Inference about Θ is then summarized in the function $f_{\Theta|U}$, which is called the posterior distribution of Θ given the data. The posterior distribution is derived from the joint distribution $f_{U,\Theta} = f_{U|\Theta} f_\Theta$ according to Bayes' formula

$$\begin{aligned} f_{\Theta|U}(\theta|u) &= \frac{f_{\Theta,U}(\theta,u)}{f_U(u)} = \frac{f_{U|\Theta}(u|\theta) f_\Theta(\theta)}{\int_\Omega f_{\Theta,U}(\theta,u)d\theta} \\ &= \frac{L(\theta;u) f_\Theta(\theta)}{\int_\Omega L(\theta;u) f_\Theta(\theta) d\theta}, \end{aligned} \tag{4.10}$$

where Ω denotes the parameter space of Θ. For brevity again the integral is used, although Θ may also be discrete. In such cases the integral should be understood as the sum. From (4.10) it is easily seen that $f_{\Theta|U}(\theta|u)$ is proportional to the likelihood multiplied by the prior; i.e.,

$$\begin{aligned} f_{\Theta|U}(\theta|u) &= c(u)^{-1} L(\theta;u) f_\Theta(\theta) \propto L(\theta;u) f_\Theta(\theta) \\ &= f_{U|\Theta}(u|\theta) f_\Theta(\theta), \end{aligned} \tag{4.11}$$

and thus involves a contribution from the observed data through $L(\theta; u)$ and a contribution from prior information quantified through $f_\Theta(\theta)$. The

[4]To make clear that the parameter θ is treated as a random variable in the Bayesian context, we use capital letters for the random variable Θ as far as possible. For parameters such as the mean μ or the covariance matrix Σ we refer to the context to see if they are taken as random or fixed.

quantity

$$c(u) = \begin{cases} \int_\Omega f_{U|\Theta}(u|\theta) f_\Theta(\theta) d\theta & \text{if } \Theta \text{ is continuous,} \\ \sum_\Omega f_{U|\Theta}(u|\theta) f_\Theta(\theta) & \text{if } \Theta \text{ is discrete,} \end{cases} \tag{4.12}$$

is usually treated as the normalizing constant of $f_{\Theta|U}(\theta|u)$ ensuring that it is a density or probability function, i.e., to integrate or sum to one. Notice that $c(u)$ is a constant when the data $U = u$ are observed. Before the data U are obtained, their distribution $f_U(u)$ is called the marginal density of U or the prior predictive distribution, which it is not conditioning on previous observations. To predict a future observation value $\widehat{u}$ when the data $U = u$ have been observed, we condition on these previous observations u. The distribution $f_{\widehat{U}|U}(\widehat{u}|u)$ of $\widehat{U}$ is called the posterior predictive distribution with

$$\begin{aligned} f_{\widehat{U}|U}(\widehat{u}|u) &= \int_\Omega f_{\widehat{U}|\Theta,U}(\widehat{u}|\theta, u) f_{\Theta|U}(\theta|u) d\theta \\ &= \int_\Omega f_{\widehat{U}|\Theta}(\widehat{u}|\theta) f_{\Theta|U}(\theta|u) d\theta \end{aligned} \tag{4.13}$$

if Θ is continuous, otherwise the sum is taken instead of the integral. Notice that usually $\widehat{U}$ and U are assumed to be conditionally independent given Θ; thus $f_{\widehat{U}|\Theta,U}(\widehat{u}|\theta, u) = f_{\widehat{U}|\Theta}(\widehat{u}|\theta)$ holds. Hence the posterior predictive distribution is conditioned on the values $U = u$ already observed and predicts a value $\widehat{U} = \widehat{u}$ that is observable.

A classical and extensive introduction to Bayesian inference is given by Box and Tiao (1992); for deeper insights into Bayesian inference and computation the interested reader is referred thereto. For further reading concerning Bayesian inference we recommend Berger (1985), Gelman et al. (1995), and Carlin and Louis (2000). Remember that we use Bayesian statistics to "solve" (i.e., replace) the problem of identifiability inherent in statistical matching by using prior information and suitable multiple imputation procedures. Posterior distributions are to be derived but no further Bayesian tools are needed. Once the different samples have been matched analysis may proceed as usual, mostly based on a frequentist viewpoint.[5] Frequentist methods, however, do not tell us what our belief in a theory should be, given the data we have actually observed. This question is usually answered by the posterior distribution $f_{\Theta|U}$. To work out this value we must first establish f_Θ; i.e., we have to formulate some "prior probability" for the theory in mind. In contrast to classical Bayesian inference we do not focus further on what we can learn about our theory given the data. Our objective is the derivation of a suitable imputation procedure that has

[5] In frequentist inference all information about unknown parameters of interest is gathered from random samples, the process of performing repetitions of an experiment, the representation of data by density or probability functions, and statistics as a function of the data. No assumptions about the parameters of interest are explicitly formulated.

good properties under the frequentist's randomization perspective and also allows us to control for the unknown conditional association of the variables never jointly observed. For the reader unfamiliar with Bayesian methods, the necessary derivations are given in detail in the appendix.

In the Bayesian framework all inference is based on a posterior density function for the unknown parameters that conditions on the quantities that are observed. Returning to our notation of Section 4.2.2 the unknown parameters are (Θ, Ξ) and the observed quantities are (U_{obs}, R). According to Bayes' theorem the posterior distribution of (Θ, Ξ) given $(U_{obs} = u_{obs}, R = r)$, i.e., the observed-data posterior distribution $f_{\Theta,\Xi|U_{obs},R} = f_{\Theta,\Xi,U_{obs},R}/f_{U_{obs},R}$, may be written as

$$\begin{aligned} f_{\Theta,\Xi|U_{obs},R}(\theta,\xi|u_{obs},r) &= c^{-1} f_{U_{obs},R|\Theta,\Xi}(u_{obs},r|\theta,\xi) f_{\Theta,\Xi}(\theta,\xi) \\ &= c^{-1} L(\theta,\xi;u_{obs},r) f_{\Theta,\Xi}(\theta,\xi) \end{aligned} \tag{4.14}$$

with normalizing constant

$$\begin{aligned} c &= \int\int f_{\Theta,U_{obs},\Theta,\Xi}(\theta,u_{obs},\theta,\xi) d\theta\, d\xi \\ &= \int\int f_{U_{obs},R|\Theta,\Xi}(u_{obs},r|\theta,\xi) f_{\Theta,\Xi}(\theta,\xi) d\theta\, d\xi\,. \end{aligned}$$

Note that $L(\theta, \xi; u_{obs}, r)$ denotes the likelihood (4.2) of the observed data considering the missingness mechanism, and $f_{\Theta,\Xi}(\theta, \xi)$ is the joint prior distribution of the parameters.

Under the assumption of MAR and the distinctness of (Θ, Ξ), which means independence of Θ and Ξ in Bayesian inference, i.e., $f_{\Theta,\Xi}(\theta,\xi) = f_\Theta(\theta) f_\Xi(\xi)$, according to (4.2) the observed-data posterior (4.14) reduces to

$$\begin{aligned} f_{\Theta,\Xi|U_{obs},R}(\theta,\xi|u_{obs},r) &= c^{-1} f_{R|U_{obs},\Theta,\Xi}(r|u_{obs},\theta,\xi) f_{U_{obs}|\Theta,\Xi}(u_{obs}|\theta,\xi) \\ &\quad f_\Theta(\theta) f_\Xi(\xi) \\ &= c^{-1} f_{R|U_{obs},\Xi}(r|u_{obs},\xi) f_{U_{obs}|\Theta}(u_{obs}|\theta) \\ &\quad f_\Theta(\theta) f_\Xi(\xi)\,. \end{aligned} \tag{4.15}$$

From the Bayesian point of view the MAR assumption requires the independence of R and Θ, i.e., $f_{R|U_{obs},\Theta,\Xi}(r|u_{obs},\theta,\xi) = f_{R|U_{obs},\Xi}(r|u_{obs},\xi)$, as well as the independence of U_{obs} and Ξ, i.e., $f_{U_{obs}|\Theta,\Xi}(u_{obs}|\theta,\xi) = f_{U_{obs}|\Theta}(u_{obs}|\theta)$ leading to (4.15) finally.

Hence the marginal posterior distribution of Θ is achieved by integrating (4.15) over the nuisance parameter Ξ with

$$\begin{aligned} f_{\Theta|U_{obs},R}(\theta|u_{obs},r) &= \int f_{\Theta,\Xi|U_{obs},R}(\theta,\xi|u_{obs},r) d\xi \\ &= \int c^{-1} f_{R|U_{obs},\Xi}(r|u_{obs},\xi) f_{U_{obs}|\Theta}(u_{obs}|\theta) \\ &\quad f_\Theta(\theta) f_\Xi(\xi) d\xi \end{aligned}$$

$$
\begin{aligned}
&= c^{-1} f_{U_{obs}|\Theta}(u_{obs}|\theta) f_\Theta(\theta) \\
&\quad \int f_{R|U_{obs},\Xi}(r|u_{obs},\xi) f_\Xi(\xi) d\xi && (4.16) \\
&\propto L(\theta; u_{obs}) f_\Theta(\theta) && (4.17) \\
&\propto f_\Theta(\theta) \prod_{i=1}^{n} \int f_{U|\Theta}(u_i|\theta) du_{mis,i}.
\end{aligned}
$$

In the matching case this function (4.17) is equivalent to

$$
\begin{aligned}
L(\theta; u_{obs}) f_\Theta(\theta) &= f_\Theta(\theta) \prod_{i=1}^{n_A} \int f_{Z,X,Y|\Theta}(z_i, x_i, y_i|\theta) dx_i \\
&\quad \prod_{i=n_A+1}^{n} \int f_{Z,X,Y|\Theta}(z_i, x_i, y_i|\theta) dy_i. && (4.18)
\end{aligned}
$$

From (4.17) we realize that $f_{\Theta|U_{obs},R}(\theta|u_{obs}, r)$ is proportional to

$$c^{-1} L(\theta; u_{obs}) f_\Theta(\theta) = f_{\Theta|U_{obs}}(\theta|u_{obs}).$$

Thus, under ignorability all information about Θ is included in the posterior that ignores the missing-data mechanism,

$$
\begin{aligned}
f_{\Theta|U_{obs}}(\theta|u_{obs}) &= c^{-1} L(\theta; u_{obs}) f_\Theta(\theta) \\
&= c^{-1} f_\Theta(\theta) \prod_{i=1}^{n} f_{U_{obs}}(u_{obs,i}; \theta). && (4.19)
\end{aligned}
$$

Again for brevity we refer to (4.19) as the "observed-data posterior" according to Schafer (1997), p. 17.

With incomplete data, however, the usual conjugate prior distributions no longer lead to posterior distributions that are recognizable or easy to summarize.

4.3 Multiple Imputation Inference

4.3.1 Introduction

The usual single imputation strategies such as mean imputation, hot deck, or regression imputation typically result in confidence intervals and p-values that ignore the uncertainty due to the missing data, because the imputed data were treated as if they were fixed known values. [6] The concept of multiple imputation (MI) reflects uncertainty in the imputation of the missing

[6] A discussion of the advantages and disadvantage of single and multiple imputation procedures may be found by the interested reader in Marker et al. (2002) and Meng (2002). Approaches for obtaining frequency valid standard errors under single imputation procedures are discussed, e.g., by Lee et al. (2002) and Shao (2002).

Unit no.	Variable						
	U_1	U_2	...	U_k	U_{k+1}	...	U_r
i	U observed				U missing		

$$U|\theta \sim f_U(u_i,\theta) \Rightarrow U_{mis}|u_{obs},\theta \sim f_{Umis|Uobs}(u_{mis,i}|u_{obs,i},\theta) \text{ for unit i e.g.}$$

$$U|\theta \sim N_r(\mu_U,\Sigma_U) \Rightarrow U_{mis}|u_{obs},\theta \sim N_{r-k}(\mu_{Umis|Uobs},\Sigma_{Umis|Uobs})$$

Figure 4.3. Example of a conditional predictive distribution

values through resulting in theoretically wider confidence intervals and thus p-values suggesting less significance than single imputation would. MI as it has been proposed by Rubin (1987) is a Monte Carlo technique replacing the missing values by $m > 1$ simulated versions, generated according to a probability distribution or, more generally, any density function indicating how likely are imputed values given the observed data. Typically m is small, with $m = 3$ or 5. Each of the imputed and thus completed data sets is first analyzed by standard methods; the results are then combined or pooled to produce estimates and confidence intervals that embed the missing data uncertainty; see Figure 4.2.

Basically, MI requires independent random draws from the posterior predictive distribution $f_{U_{mis}|U_{obs}}$ of the missing data given the observed data. Therefore we usually perform

(1) random draws of the parameters Θ according to their observed-data posterior distribution $f_{\Theta|U_{obs}}$ as well as

(2) random draws of U_{mis} according to their conditional predictive distribution
$f_{U_{mis}|U_{obs},\Theta}$.

Because

$$f_{U_{mis}|U_{obs}}(u_{mis}|u_{obs}) = \int f_{U_{mis}|U_{obs},\Theta}(u_{mis}|u_{obs},\theta) f_{\Theta|U_{obs}}(\theta|u_{obs}) d\theta \tag{4.20}$$

holds, analogous to (4.13), with (1) and (2) we achieve imputations of U_{mis} from their posterior predictive distribution $f_{U_{mis}|U_{obs}}$, because it is equivalent to the conditional predictive distribution $f_{U_{mis}|U_{obs},\Theta}$ averaged over the observed-data posterior $f_{\Theta|U_{obs}}(\theta|u_{obs})$ of Θ. For many models the conditional predictive distribution $f_{U_{mis}|U_{obs},\Theta}$ is rather straightforward due to the data model used; see as an example Figure 4.3. It often may easily be formulated for each unit with missing data.

On the contrary the corresponding observed-data posteriors $f_{\Theta|U_{obs}}$ usually are difficult to derive for those units with missing data, especially when the data have a multivariate structure and different missing data patterns.

The observed-data posteriors are often not standard distributions from which random numbers can easily be generated. Imputing missing values according to the posterior predictive distribution $f_{U_{mis}|U_{obs}}$ may in general be difficult. However, simpler methods have been developed to enable multiple imputation on the grounds of Markov chain Monte Carlo (MCMC) techniques; they are extensively discussed by Schafer (1997). In MCMC the desired distributions $f_{U_{mis}|U_{obs}}$ and $f_{\Theta|U_{obs}}$ are achieved as stationary distributions of Markov chains which are based on the easier to compute complete-data distributions.

As we have previously mentioned it can be quite difficult to prove that an imputation procedure is proper in the sense of Rubin (1987), pp. 118–119. Usually it is to be expected that imputations which are independently drawn from (4.20) should be proper or at least approximately proper. However, whether the imputations really lead to frequency valid answers is left to be proven for each imputation procedure. Therefore Schafer (1997), p. 105, coined the term Bayesianly proper for an imputation procedure that generates independent realizations of the posterior predictive distribution $f_{U_{mis}|U_{obs}}$. Thus Bayesianly proper imputations do not necessarily imply proper imputations. For example, the multiple imputation method based on a regression model proposed by Schenker and Welsh (1988) is Bayesianly proper, but not proper in Rubin's frequentist sense. But, once a procedure is Bayesianly proper, at least some of the conditions of being proper are automatically satisfied. In Section 4.3.3 we take up this discussion again.

4.3.2 The Multiple Imputation Principle

Let us for the moment only consider the data U to be random and let θ denote a scalar quantity of interest that is to be estimated, such as a mean, variance, or correlation coefficient. As described before, $U = (U_{obs}, U_{mis})$ denotes the random variables concerning the data with observed and missing parts, and $\widehat{\theta} = \widehat{\theta}(U)$ denotes the statistic that would be used to estimate θ if the data were complete. In our matching case a generic estimation task is given with the correlation coefficient $\theta = \rho_{XY}$ of the variables never jointly observed usually estimated by the sample correlation coefficient, for example, Pearson's product moment correlation coefficient $\widehat{\theta} = \widehat{\rho}_{XY} = (\sum_{i=1}^{N}(X_i - \overline{X})(Y_i - \overline{Y}))/(\sqrt{\sum_{k=1}^{N}(X_k - \overline{X})^2 \sum_{k=1}^{N}(Y_k - \overline{Y})^2})$. Furthermore, let $\widehat{var}(\widehat{\theta}) = \widehat{var}(\widehat{\theta}(U))$ be the variance estimate of $\widehat{\theta}(U)$ based on the complete data set.

The MI principle as proposed by Rubin (1987) assumes that $\widehat{\theta}$ and $\widehat{var}(\widehat{\theta})$ can be regarded as an approximate complete-data posterior mean and variance for θ, with $\widehat{\theta} \approx E(\Theta|u_{obs}, u_{mis})$ and $\widehat{var}(\widehat{\theta}) \approx var(\Theta|u_{obs}, u_{mis})$ based on a suitable complete-data model and prior; see also Schafer (1997), p. 108. Moreover, we must assume that with complete data, tests and interval estimates based on the normal approximation

$$(\widehat{\theta} - \theta)/\sqrt{\widehat{var}(\widehat{\theta})} \sim N_1(0,1) \tag{4.21}$$

should work well. Hence, we assume that the complete-data inference can be based on

$$\widehat{\theta} \sim N_1(\theta, var(\widehat{\theta}))$$

and that $\widehat{var}(\widehat{\theta})$ is of lower-order variability than $var(\widehat{\theta})$; see Li et al. (1991). Notice that the usual maximum-likelihood estimates and their asymptotic variances derived from the inverted Fisher information matrix typically satisfy these assumptions. Sometimes it is necessary to transform the estimate $\widehat{\theta}$ to a scale for which the normal approximation can be applied. For example, we can use the so-called z-transformation for the correlation coefficient estimate, with $z(\widehat{\rho}) = (1/2)\ln((1+\widehat{\rho})/(1-\widehat{\rho}))$, which makes $z(\widehat{\rho})$ approximately normally distributed with mean $z(\rho)$ and constant variance $1/(n-3)$; see Schafer (1997), p. 216 and Brand (1999), p. 116.

Suppose now that the data are missing and we make $m > 1$ independent simulated imputations $(U_{obs}, U_{mis}^{(1)})$, $(U_{obs}, U_{mis}^{(2)})$, ..., $(U_{obs}, U_{mis}^{(m)})$ enabling us to calculate the imputed data estimate $\widehat{\theta}^{(t)} = \widehat{\theta}(U_{obs}, U_{mis}^{(t)})$ along with its estimated variance $\widehat{var}(\widehat{\theta}^{(t)}) = \widehat{var}(\widehat{\theta}(U_{obs}, U_{mis}^{(t)}))$, $t = 1, 2, \ldots, m$. Figure 4.4 illustrates the multiple imputation principle. From these m imputed data sets the multiple imputation estimates are computed.

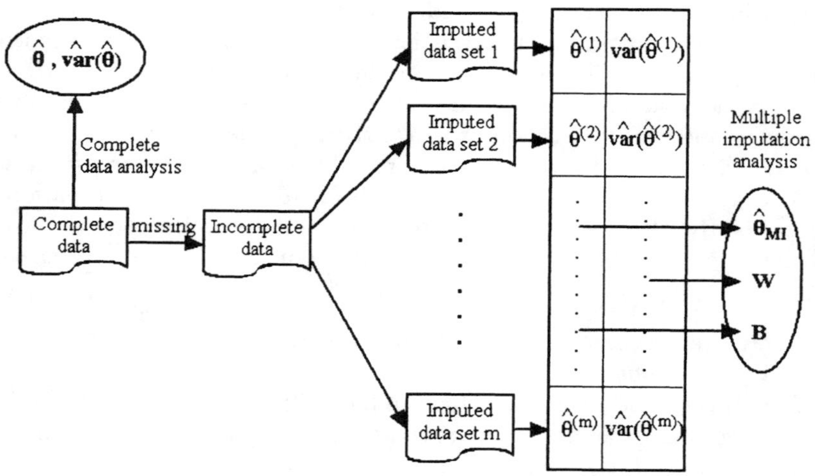

Figure 4.4. The multiple imputation principle

The MI point estimate for θ is simply the average

$$\widehat{\theta}_{MI} = \frac{1}{m}\sum_{t=1}^{m}\widehat{\theta}^{(t)}. \tag{4.22}$$

To obtain a standard error $\sqrt{\widehat{var}(\widehat{\theta}_{MI})}$ for the MI estimate $\widehat{\theta}_{MI}$ we first calculate the "between-imputation" variance

$$\widehat{var}(\widehat{\theta})_{between} = B = \frac{1}{m-1}\sum_{t=1}^{m}(\widehat{\theta}^{(t)} - \widehat{\theta}_{MI})^2, \tag{4.23}$$

and then the "within-imputation" variance

$$\widehat{var}(\widehat{\theta})_{within} = W = \frac{1}{m}\sum_{t=1}^{m}\widehat{var}(\widehat{\theta}^{(t)}). \tag{4.24}$$

Finally, the estimated total variance is defined by

$$\begin{aligned} \widehat{var}(\widehat{\theta}_{MI}) &= T = \widehat{var}(\widehat{\theta})_{within} + (1 + \frac{1}{m})\widehat{var}(\widehat{\theta})_{between} \\ &= W + \frac{m+1}{m}B. \end{aligned} \tag{4.25}$$

Notice that the term $((m+1)/m)B$ enlarges the total variance estimate T compared to the usual analysis of variance with $T = B + W$. Tests and two-sided $(1-\alpha)100\%$ interval estimates may be based on the Student's t-distribution

$$(\widehat{\theta}_{MI} - \theta)/\sqrt{T} \sim t_v \text{ and } \widehat{\theta}_{MI} \pm t_{v,1-\alpha/2}\sqrt{T} \tag{4.26}$$

with degrees of freedom

$$v = (m-1)\left(1 + \frac{W}{(1+m^{-1})B}\right)^2; \tag{4.27}$$

see Rubin (1987), pp. 76–77. Thus we realize that the multiple imputation interval estimate is expected to produce a larger interval than an estimate based only on the observed cases or based only on one single imputation. The multiple imputation interval estimates are widened to account for the missing data uncertainty and simulation error; see Schafer (1999a).

4.3.3 Efficiency of the Multiple Imputation Estimates

Rubin (1987), p. 114, shows that the relative efficiency of an estimate based on m imputations to one based on $m = \infty$ number of imputations is approximately $1 + \lambda/m$ to 1, where λ is the rate of missing information. Taking $m = 3$ multiple imputations and assuming a fraction of 50% missing information an estimate based on this $m = 3$ imputations has a standard error that is about 8% higher than one based on $m = \infty$, because

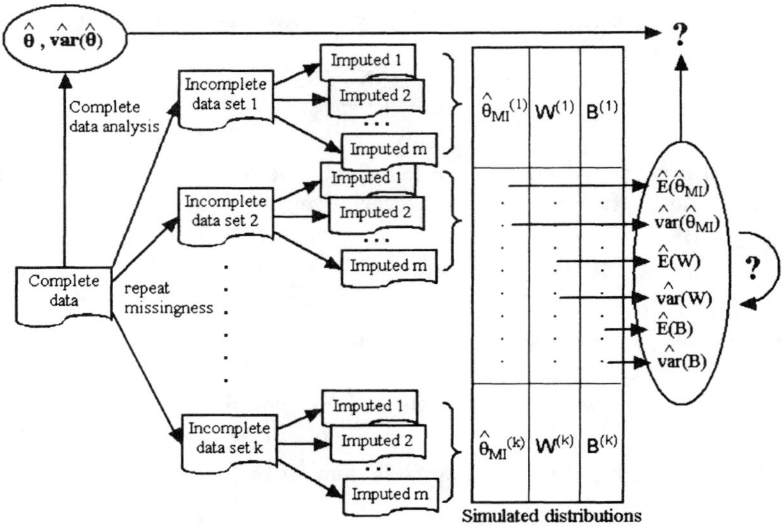

Figure 4.5. Simulation process to verify a proper imputation method

$\sqrt{1 + 0.5/3} = 1.0801$. Schafer (1999a) states that unless the fraction of missing information is unusually high (i.e., far more than 50%), there is little benefit in using more than 5 to 10 imputations.

We do not want to replicate all the inferential questions and justifications for the MI principle in general here; they are extensively described by Rubin (1987, 1996), Schafer (1997), and Brand (1999). In particular we discuss the properties of our alternative approaches to statistical matching derived in the following sections on the basis of a simulation study at this chapter's end. An illustration of the verification of a proper imputation method is shown in Figure 4.5 which is adopted from Brand (1999), p. 115, and extended. Notice that the concept of proper imputation methods requires the complete-data inference to be "randomization-valid"; i.e., for repeated sampling from the underlying population the following should approximately hold,

$$\widehat{\theta} \sim N_1(\theta, var(\widehat{\theta})), \qquad E(\widehat{var}(\widehat{\theta})) = var(\widehat{\theta}),$$

and $\widehat{var}(\widehat{\theta})$ has less variability than $\widehat{\theta}$. See Rubin (1987), p. 118.

To evaluate whether a MI technique is proper for a set of complete-data statistics $(\widehat{\theta}(U), \widehat{var}(\widehat{\theta}(U)))$ by means of Monte Carlo simulation the following simplified validity conditions may be discussed.

$$\begin{aligned} E(\widehat{\theta}_{MI}) &= \widehat{\theta}(U), \\ E(W) &= \widehat{var}(\widehat{\theta}(U)), \\ E(B) &= var(\widehat{\theta}_{MI}). \end{aligned} \tag{4.28}$$

As the number of imputations becomes large (i.e., $m \to \infty$,) these equations (4.28) demand that $\widehat{\theta}_{MI}$, W, and B are unbiased estimates of the statistics $\widehat{\theta}(U)$, $\widehat{var}(\widehat{\theta}(U))$, and $var(\widehat{\theta}_{MI})$. Notice that $\widehat{\theta}(U)$ and $\widehat{var}(\widehat{\theta}(U))$ are based on the hypothetical complete data set when no data are missing and $var(\widehat{\theta}_{MI})$ describes the true variance of the MI estimate for a given observed data set and $m = \infty$ number of imputations.

More precisely, it is required that $(\widehat{\theta}_{MI} - \widehat{\theta}(U))/\sqrt{B}$ should become approximately $N(0,1)$, when the number of imputations gets large and the data U are held fixed. The within-imputation variance W should be an unbiased estimate of the complete-data variance estimate $\widehat{var}(\widehat{\theta}(U))$, and the variances of W and B should be smaller than the true variance $var(\widehat{\theta}_{MI})$ of the MI estimate, if again the data are regarded as fixed and $m = \infty$. Finally, over repeated samples the true variance $var(\widehat{\theta}_{MI})$ should be of lower variability than $\widehat{\theta}(U)$.[7] By means of Monte Carlo simulations we may recognize that an imputation procedure can be (more or less) proper for a set of complete-data statistics $(\widehat{\theta}(U), \widehat{var}(\widehat{\theta}(U)))$ by verifying whether

$$\begin{aligned} \widehat{E}(\widehat{\theta}_{MI}) &\approx \widehat{\theta}(U), \\ \widehat{E}(W) &\approx \widehat{var}(\widehat{\theta}(U)), \\ \widehat{var}(\widehat{\theta}_{MI}) &\approx (1+m^{-1})\widehat{E}(B) \end{aligned} \tag{4.29}$$

holds; see Brand (1999), pp. 114–117. A schematic overview of the simulation process to verify whether an imputation method may be proper is presented in Figure 4.5. We refer to these approximations (4.30) in our simulation study at the end of this chapter.

4.3.4 Normally Distributed Data Not Jointly Observed

As proposed by Rubin (1987), pp. 187–188, we derive a Bayesianly proper multiple imputation technique for the simple matching case with two variables X and Y not jointly observed.

To guarantee Bayesianly proper multiple imputations we first perform random draws for the parameters from their observed-data posterior. Then random draws for the missing values are performed following their conditional predictive distribution. If we assume the noninformative prior with

$$f_{\mu,\Sigma}(\mu,\Sigma) \approx f_\mu(\mu) f_\Sigma(\Sigma) = f_\mu(\mu) f_{\sigma_X^2,\sigma_Y^2|\rho}(\sigma_X^2,\sigma_Y^2|\rho) f_\rho(\rho) \propto \sigma_X^{-2}\sigma_Y^{-2} f_\rho(\rho),$$

[7] For the original definition of a proper imputation method see Rubin (1987), pp. 118–119; a summary is given by Schafer (1997), pp. 144–145. A detailed discussion of the verification of proper multiple imputations is provided by Brand (1999), pp. 75–91 and pp. 114–117.

we find the joint posterior distribution given by the following marginal and conditional posterior distributions.

$$\begin{aligned} \mu_Y|\sigma_Y^2, y &\sim N(\widehat{\mu}_Y, \frac{\sigma_Y^2}{n_A}), \qquad \sigma_Y^2|y \sim \frac{\sum_{i=1}^{n_A}(y_i - \widehat{\mu}_Y)^2}{\chi^2_{(n_A-1)}} = \chi^{-2}_{(n_A-1)} s_Y^2, \\ \mu_X|\sigma_X^2, x &\sim N(\widehat{\mu}_X, \frac{\sigma_X^2}{n_B}), \qquad \sigma_X^2|x \sim \frac{\sum_{i=1}^{n_B}(x_i - \widehat{\mu}_X)^2}{\chi^2_{(n_B-1)}} = \chi^{-2}_{(n_B-1)} s_X^2, \\ \rho &\sim f_\rho, \end{aligned}$$

with $\widehat{\mu}_X = \overline{x} = (1/n_B)\sum_{i=1}^{n_B} x_i$, $s_X^2 = \sum_{i=1}^{n_B}(x_i - \overline{x})^2$, $\widehat{\mu}_Y = \overline{y} = (1/n_A)\sum_{i=1}^{n_A} y_i$, and $s_Y^2 = \sum_{i=1}^{n_A}(y_i - \overline{y})^2$. Since the posterior distribution of ρ equals its prior distribution, we may, for example, assume ρ to be uniformly distributed over $]-1,1[$ or take some arbitrary values for ρ such as -0.9, 0, or 0.9.

Having thus drawn actual values $\mu^{(t)}$ and $\Sigma^{(t)}$ for the parameters μ and Σ from their observed-data posterior distribution, random values for the missing data $X^{(t)} = (X_1^{(t)}, \dots, X_{n_A}^{(t)})$ and $Y^{(t)} = (Y_{n_A+1}^{(t)}, \dots, Y_n^{(t)})$ are generated according to their conditional predictive distribution $f_{U_{mis}|U_{obs},\Theta^{(t)}}$, i.e., $f_{X|Y,\mu^{(t)},\Sigma^{(t)}}$ in file A and $f_{Y|X,\mu^{(t)},\Sigma^{(t)}}$ in file B, respectively. In particular we draw for each unit for file A

$$X_i^{(t)}|y_i, \mu^{(t)}, \Sigma^{(t)} \sim N_1\left(\mu_X^{(t)} + \rho^{(t)}\frac{\sigma_X^{(t)}}{\sigma_Y^{(t)}}(y_i - \mu_Y^{(t)}); \sigma_{XX}^{(t)}(1-\rho^{2(t)})\right), \quad \text{for } i = 1, 2, \dots, n_A, \tag{4.30}$$

and for file B

$$Y_i^{(t)}|x_i, \mu^{(t)}, \Sigma^{(t)} \sim N_1\left(\mu_Y^{(t)} + \rho^{(t)}\frac{\sigma_Y^{(t)}}{\sigma_X^{(t)}}(x_i - \mu_X^{(t)}); \sigma_{YY}^{(t)}(1-\rho^{2(t)})\right), \quad \text{for } i = n_A + 1, \dots, n, \tag{4.31}$$

assuming that the imputations are independent of each other for each unit. To abbreviate the notation we often describe the imputation procedure for the whole vector at once instead of (4.31). For example, the imputation of Y follows

$$Y^{(t)}|x, \mu^{(t)}, \Sigma^{(t)} \sim N_{n_B}\left(\mu_Y^{(t)} + \rho^{(t)}\frac{\sigma_Y^{(t)}}{\sigma_X^{(t)}}(x - \mu_X^{(t)}); \sigma_{YY}^{(t)}(1-\rho^{2(t)})I_{n_B}\right)$$

with $x = (x_1, x_2, \dots, x_{n_A})'$, indicating by means of the identity matrix I_{n_B} that the imputations are independently generated for each unit i of file B. Furthermore, we sometimes write $\sigma_Y^2 = \sigma_{YY}, \sigma_X^2 = \sigma_{XX}$ to ease reading, especially when superscripts are used. Each new draw for the parameters generates new draws for the missing values and thus generates a new completed data set.

Taking up the small example from Table 4.1 using only the variables X and Y we may generate $m = 3$ completed data sets listed in Table 4.4

Original x_i	Original y_i	First imputation x_i	First imputation y_i	Second imputation x_i	Second imputation y_i	Third imputation x_i	Third imputation y_i
NA	10	-1.41	10.00	8.57	10.00	11.87	10.00
NA	14	18.35	14.00	6.53	14.00	9.11	14.00
NA	16	5.01	16.00	0.05	16.00	7.05	16.00
NA	15	15.35	15.00	1.30	15.00	8.18	15.00
NA	20	17.64	20.00	10.02	20.00	8.46	20.00
NA	4	18.02	4.00	-1.03	4.00	10.18	4.00
1	NA	1.00	17.02	1.00	14.21	1.00	8.04
5	NA	5.00	10.27	5.00	18.60	5.00	11.20
8	NA	8.00	25.21	8.00	19.40	8.00	12.30
12	NA	12.00	14.11	12.00	13.41	12.00	11.05
$\widehat{\sigma}_{XY}^{(t)}$		-2.3529		7.3066		0.3674	
$\widehat{\rho}_{XY}^{(t)}$		-0.0611		0.3711		0.0281	
$z(\widehat{\rho}_{XY}^{(t)})$		-0.0612		0.3897		0.0281	

Table 4.4. Arbitrary and imputed bivariate data set

according to the MI routine described above. From the observed data we get $\widehat{\mu}_X = 6.5, s_X^2 = 65, \widehat{\mu}_Y = 13.1667$, and $s_Y^2 = 152.8333$. As a prior value of ρ we arbitrarily choose $\rho = 0$.

Notice that under a large sample size the maximum likelihood covariance estimate $\widehat{\sigma}_{XY} = (1/n)\sum_{i=1}^{n}(x_i - \overline{x})(y_i - \overline{y})$ is approximately normally distributed and the multiple imputation estimate $\widehat{\sigma}_{MI}$ could be derived directly with $\widehat{\sigma}_{MI} = (1/m)\sum_{t=1}^{m}\widehat{\sigma}_{XY}^{(t)}$. Clearly in this example no asymptotic holds, thus we get the multiple imputation estimate of ρ by means of the z-transformed estimates

$$z(\widehat{\rho}^{(t)}) = \frac{1}{2}\ln\frac{1+\widehat{\rho}_{XY}^{(t)}}{1-\widehat{\rho}_{XY}^{(t)}},$$

with $z(\widehat{\rho}^{(t)})$ being approximately unbiased and normally distributed with a constant variance $1/(n-3)$; see Stuart and Ord (1991), p. 984. We calculate

$$z(\widehat{\rho})_{MI} = \frac{1}{m}\sum_{t=1}^{m}\frac{1}{2}\ln\frac{1+\widehat{\rho}_{XY}^{(t)}}{1-\widehat{\rho}_{XY}^{(t)}} = \frac{1}{6}(-0.0612+0.3897+0.0281) = 0.1189.$$

By back-transformation we achieve the MI estimate $\widehat{\rho}_{MI} = (e^{2\cdot 0.1189} - 1)/(e^{2\cdot 0.1189}+1) = 0.1183$, which is not very different from directly averaging $(1/m)\sum_{t=1}^{m}\widehat{\rho}_{XY}^{(t)} = 0.1127$, because $z(\widehat{\rho}) \approx \widehat{\rho}$ for values of $\widehat{\rho}$ near zero.

To proceed further, MI inference about ρ is derived by calculating

- the within-imputation variance, which is constant for the z-transformation with $W = 1/7$, because $var(z(\widehat{\rho})) = 1/(10-3)$;

- the between-imputation variance, $B = (1/(m-1))\sum_{t=1}^{m}(z(\widehat{\rho}_{XY}^{(t)}) - z(\widehat{\rho})_{MI})^2 = 0.0570$;
- the total variance is $T = W + (1 + 1/3)B = 0.2189$;
- the 95% MI confidence interval for $z(\rho)$ is given by $z(\widehat{\rho})_{MI} \pm t_{v,0.975}\sqrt{T}$ with $v = 16.5853$; thus we get $0.1189 \pm 2.1138\sqrt{0.2189}$, i.e., $[-0.8701, 1.1079]$;
- the back-transformed confidence interval for ρ finally is given by $[-0.7014, 0.8033]$ for this small data example.

4.3.5 Discussion

If we perform more imputations with different prior values for ρ we will always reproduce more or less the prior value of ρ by the corresponding MI estimate. Contrary to the EM algorithm, we are now able to control for the results induced by multiple imputation due to the prior distribution. Since statistical matching without common variables is only of interest for demonstration purposes, in the following sections we investigate how the common variables may be used efficiently to improve the estimation procedure. Different data models to incorporate the common variables in the imputation procedure are discussed. By means of multiple imputation and the efficient use of the common variables we estimate the bounds of the unconditional association of the variables never jointly observed.

4.4 Regression Imputation with Random Residuals

4.4.1 Introduction

Some intuitively appealing imputation approaches are based on (frequentist) regression imputation. The predictive mean matching, for instance, as introduced by Rubin (1986) and extended by Little (1988) to any pattern of missingness allowing multiple imputations has become fairly widespread.[8] Instead of imputing predicted or observed values from suitable donor units sometimes a random residual is added to account for the so-called regression to the mean. The addition of such a random residual extends the conventional single regression imputation procedure to produce multiple imputations. However, such procedures not derived within the Bayesian framework are often not proper. Although a random residual is added, the regression imputation is said to be an improper procedure; see Schafer (1999a).

[8]See some overviews about recent imputation techniques by Nordholt (1998) or Rässler (2000).

Nevertheless, Rubin (1986) proposed an implicit model based on a predictive mean matching regression approach in the context of statistical matching. A similar procedure for assessing the effect of alternative assumptions of the inestimable unconditional association of the variables not jointly observed was published by Kadane in 1978 which is now reprinted; see Kadane (2001). Moriarity and Scheuren (2001) give an extension of Kadane's approach with better properties.

We discuss Rubin's (1986) procedure here, especially because the following noniterative Bayesian regression model can be viewed as the Bayesian version of this approach. Moreover, the method is extended by adding a random residual. In the final simulation study at this chapter's end it is shown that this frequentist imputation procedure will not yield proper multiple imputations.

4.4.2 Imputation Procedure

A regression imputation procedure for the matching case can be constructed as follows. Assume the general linear model for both data sets with

$$\begin{aligned} (\text{file } A) \quad Y &= Z_A \beta_{YZ} + U_A, \\ (\text{file } B) \quad X &= Z_B \beta_{XZ} + U_B, \end{aligned} \tag{4.32}$$

with Z_A $n_A \times k$ and Z_B $n_B \times k$ matrices of known values each with rank k. Usually we treat Z as the common derivative matrix including the constant $(1, 1, \ldots, 1)'$; thus we let z_1 describe the constant. Y and X correspond to any multivariate variables as pictured in Figure 4.1. Thus X denotes a $n_B \times q$ matrix and Y a $n_A \times p$ matrix according to the general multivariate normal model; see Box and Tiao (1992), pp. 423–425. Notice that we denote a matrix with capital letters as it is common practice in statistical theory, although it sometimes does not refer to a matrix of random variables. On the contrary we write all variables, denoted by matrices or not, with small letters when a condition is described (but not when used as indices). Thus consistency with the former sections is preserved and the reader is not bothered with an "inflated" notation. One should further notice that although the common variables Z are usually regarded as being fixed, in a slight abuse of notation we write $\sigma^2_{Y|Z}$ instead of σ^2_Y and so on just to correspond to the distinction between the unconditional association ρ_{XY} of X and Y and the conditional association $\rho_{XY|Z}$ of X and Y given $Z = z$.

We assume a multivariate normal data model for $(X, Y|Z = z) = (X_1, X_2, \ldots, X_q, Y_1, Y_2, \ldots, Y_p|Z = z)$ with a given common variable $Z = z$ and the parameters are also given; no Bayesian inference is used here. The expectation is $\mu_{XY|Z}$ and the covariance matrix $\Sigma_{XY|Z}$ is denoted by

$$\Sigma_{XY|Z} = \left(\begin{array}{ccc|ccc} \sigma_{X_1X_1|Z} & \cdots & \sigma_{X_1X_q|Z} & \sigma_{X_1Y_1|Z} & \cdots & \sigma_{X_1Y_p|Z} \\ \cdots & \cdots & \cdots & \cdots & \cdots & \cdots \\ \sigma_{X_qX_1|Z} & \cdots & \sigma_{X_qX_q|Z} & \sigma_{X_qY_1|Z} & \cdots & \sigma_{X_qY_p|Z} \\ \hline \sigma_{Y_1X_1|Z} & \cdots & \sigma_{Y_1X_q|Z} & \sigma_{Y_1Y_1|Z} & \cdots & \sigma_{Y_1Y_p|Z} \\ \cdots & \cdots & \cdots & \cdots & \cdots & \cdots \\ \sigma_{Y_pX_1|Z} & \cdots & \sigma_{Y_pX_q|Z} & \sigma_{Y_pY_1|Z} & \cdots & \sigma_{Y_pY_p|Z} \end{array}\right) = \begin{pmatrix} \Sigma_{11} & \Sigma_{12} \\ \Sigma_{21} & \Sigma_{22} \end{pmatrix}; \qquad (4.33)$$

then the residual $U_A \sim N_{pn_A}(0, \Sigma_{22} \otimes I_{n_A})$ and $U_B \sim N_{qn_B}(0, \Sigma_{11} \otimes I_{n_B})$. If X and Y only describe univariate variables, then $p = q = 1$ with $U_A \sim N_{n_A}(0, \sigma_A^2 I_{n_A})$ as well as $U_B \sim N_{n_B}(0, \sigma_B^2 I_{n_B})$ holds. Again this data model assumes that the units can be observed independently for $i = 1, 2, \ldots, n$. The correlation structure refers to the variables

$$X_{1i}, X_{2i}, \ldots, X_{qi}, Y_{1i}, Y_{2i}, \ldots, Y_{pi}$$

for each unit $i = 1, 2, \ldots, n$. For abbreviation we use the Kronecker product $\otimes$ denoting that the variables X_i and Y_i of each unit i, $i = 1, 2, \ldots, n$, are correlated but no correlation of the variables is assumed between the units.

The maximum likelihood (or ordinary least squares) estimates derived from the data model are

$$\widehat{\beta}_{YZ} = (Z_A' Z_A)^{-1} Z_A' Y \quad \text{and} \quad \widehat{\beta}_{XZ} = (Z_B' Z_B)^{-1} Z_B' X .$$

With the multivariate X and Y we get a $k \times p$ matrix of parameters β_{YZ} as well as a $k \times q$ matrix of parameters β_{XZ}.

Using the estimates $\widehat{\beta}_{YZ}$ and $\widehat{\beta}_{XZ}$ the residual matrices Σ_{11} and Σ_{22} can be estimated for each regression with

$$\begin{aligned} S_Y/(n_A - k) &= (Y - Z_A\widehat{\beta}_{YZ})'(Y - Z_A\widehat{\beta}_{YZ})/(n_A - k) \\ &= \widehat{\Sigma}_{22} = \widehat{\Sigma}_{Y|Z} = \{\widehat{\sigma}_{Y_iY_j|Z}\},\ i,j = 1,2,\ldots,p \\ S_X/(n_B - k) &= (X - Z_B\widehat{\beta}_{XZ})'(X - Z_B\widehat{\beta}_{XZ})/(n_B - k) \\ &= \widehat{\Sigma}_{11} = \widehat{\Sigma}_{X|Z} = \{\widehat{\sigma}_{X_iX_j|Z}\},\ i,j = 1,2,\ldots,q. \end{aligned} \qquad (4.34)$$

Now we refer to the linear regression of X on Z and Y for file A and Y on Z and X for file B, respectively. Thus we model

$$\begin{aligned} (\text{file } A) \quad X &= Z_A\beta_{XZ.Y} + Y\beta_{XY.Z} + V_A, \\ & \quad V_A \sim N_{n_Aq}(0, \Sigma_{X|ZY} \otimes I_{n_A}), \\ (\text{file } B) \quad Y &= Z_B\beta_{YZ.X} + X\beta_{YX.Z} + V_B, \\ & \quad V_B \sim N_{n_Bp}(0, \Sigma_{Y|ZX} \otimes I_{n_B}) . \end{aligned} \qquad (4.35)$$

To estimate the parameters of (4.35) we first calculate the conditional covariance matrix $\widehat{\Sigma}_{XY|Z}$. Therefore any "prior" information may be used to fix the conditional correlation matrix $R_{XY|Z}$,

$$R_{XY|Z} = \begin{pmatrix} \rho_{X_1Y_1|Z} & \cdots & \rho_{X_1Y_p|Z} \\ \cdots & \cdots & \cdots \\ \rho_{X_qY_1|Z} & \cdots & \rho_{X_qY_p|Z} \end{pmatrix} = \{\rho_{X_iY_j|Z}\}, \tag{4.36}$$

for $i = 1, 2, \ldots, q, j = 1, 2, \ldots, p$ denoting a matrix of size $q \times p$.[9] If no information is available it may be convenient to choose $R_{XY|Z} = 0_{q\times p}$. Finally, the estimate of Σ_{12} is calculated by means of (4.34) with

$$\begin{aligned} \widehat{\Sigma}_{12} &= \left\{\rho_{X_iY_j|Z}\sqrt{\widehat{\sigma}_{X_iX_i|Z}\widehat{\sigma}_{Y_jY_j|Z}}\right\}, \qquad (4.37) \\ &\quad i = 1, 2, \ldots, q, j = 1, 2, \ldots, p, \text{ and} \\ \widehat{\Sigma}_{XY|Z} &= \begin{pmatrix} \widehat{\Sigma}_{X|Z} & \widehat{\Sigma}_{12} \\ \widehat{\Sigma}_{21} & \widehat{\Sigma}_{Y|Z} \end{pmatrix}, \quad \widehat{\Sigma}_{21} = \widehat{\Sigma}'_{12}. \qquad (4.38) \end{aligned}$$

The regression parameters for (4.35) are derived according to Cox and Wermuth (1998), p. 69. In the univariate cases with $p = q = 1$ we get from $\widehat{\sigma}_{12} = \rho_{XY|Z}\sqrt{\widehat{\sigma}_{X|Z}\widehat{\sigma}_{Y|Z}}$ the parameter estimate $\widehat{\beta}_{XY.Z} = \widehat{\sigma}_{12}/\widehat{\sigma}^2_{Y|Z}$ for file A and $\widehat{\beta}_{YX.Z} = \widehat{\sigma}_{12}/\widehat{\sigma}^2_{X|Z}$ for file B.

In the multivariate regression model we calculate

$$\begin{aligned} &(\text{file } A) \quad \widehat{\beta}_{XZ.Y} = \widehat{\beta}_{XZ} - \widehat{\beta}_{YZ}\widehat{\beta}_{XY.Z} \text{ with } \widehat{\beta}_{XY.Z} = \widehat{\Sigma}^{-1}_{Y|Z}\widehat{\Sigma}_{21}, \text{ and} \\ &(\text{file } B) \quad \widehat{\beta}_{YZ.X} = \widehat{\beta}_{YZ} - \widehat{\beta}_{XZ}\widehat{\beta}_{YX.Z} \text{ with } \widehat{\beta}_{YX.Z} = \widehat{\Sigma}^{-1}_{X|Z}\widehat{\Sigma}_{12}. \qquad (4.39) \end{aligned}$$

The predicted regression values of all X and Y variables are given by

$$\begin{aligned} (\text{file } A) \quad \widehat{X} &= \widehat{X}_A = Z_A\widehat{\beta}_{XZ.Y} + Y\widehat{\beta}_{XY.Z}, \\ \widehat{Y} &= \widehat{Y}_A = Z_A\widehat{\beta}_{YZ.X} + \widehat{X}_A\widehat{\beta}_{YX.Z}, \\ (\text{file } B) \quad \widehat{Y} &= \widehat{Y}_B = Z_B\widehat{\beta}_{YZ.X} + X\widehat{\beta}_{YX.Z}, \text{ and} \\ \widehat{X} &= \widehat{X}_B = Z_B\widehat{\beta}_{XZ.Y} + \widehat{Y}_B\widehat{\beta}_{XY.Z}. \qquad (4.40) \end{aligned}$$

The residual variances $\Sigma_{X|ZY}$ and $\Sigma_{Y|ZX}$ of the regression (4.35) can now be estimated using (4.40):

$$\begin{aligned} \widehat{\Sigma}_{Y|ZX} &= (Y - \widehat{Y})'(Y - \widehat{Y})/(n_A - (k+q)) \\ &\quad \text{with } Y \text{ and } \widehat{Y} = \widehat{Y}_A \text{ taken from file } A \\ \widehat{\Sigma}_{X|ZY} &= (X - \widehat{X})'(X - \widehat{X})/(n_B - (k+p)) \\ &\quad \text{with } X \text{ and } \widehat{X} = \widehat{X}_B \text{ taken from file } B. \end{aligned}$$

---

[9]Notice that this regression imputation method is not based on Bayesian inference. Thus, prior information refers to any arbitrarily chosen values for the conditional correlation of X and Y given $Z = z$ which is not estimable from the observed data.

In the univariate case we compute

$$\widehat{\sigma}^2_{Y|ZX} = \frac{1}{n_A - k - 1}\sum_{i=1}^{n_A}(y_i - \widehat{y}_i)^2 \text{ and } \widehat{\sigma}^2_{X|ZY} = \frac{1}{n_B - k - 1}\sum_{i=1}^{n_B}(x_i - \widehat{x}_i)^2.$$

Finally, random residuals can be generated according to $\widehat{V}_{A,i} \sim N_q(0, \widehat{\Sigma}_{X|ZY})$ for each single row $i = 1, 2, \ldots, n_A$ and $\widehat{V}_{B,i} \sim N_p(0, \widehat{\Sigma}_{Y|ZX})$ for each single row $i = 1, 2, \ldots, n_B$. More generally, we write

$$\begin{aligned}\widehat{V}_A &\sim N_{qn_A}(0, \widehat{\Sigma}_{X|ZY} \otimes I_{nA})\\ \widehat{V}_B &\sim N_{pn_B}(0, \widehat{\Sigma}_{Y|ZX} \otimes I_{nB}).\end{aligned}$$

These randomly generated values are added to the regression output; finally the imputed values are

$$\begin{aligned}\text{(file } A) \quad \widehat{X} &= Z_A\widehat{\beta}_{XZ.Y} + Y\widehat{\beta}_{XY.Z} + \widehat{V}_A\,,\\ \text{(file } B) \quad \widehat{Y} &= Z_B\widehat{\beta}_{YZ.X} + X\widehat{\beta}_{YX.Z} + \widehat{V}_B\,.\end{aligned} \tag{4.41}$$

The calculation of the imputed values according to (4.41) is equivalent to drawing the missing values from their conditional predictive distribution $f_{U_{mis}|U_{obs},\Theta}(u_{mis}|u_{obs}, \widehat{\theta})$ given the observed data and some actual parameter values $\widehat{\theta}$; i.e.,

$$\begin{aligned}\text{(file } A) \quad \widehat{X}|y &\sim N_{qn_A}(\widehat{\mu}_{X|ZY}, \widehat{\Sigma}_{X|ZY} \otimes I_{nA}) \quad \text{and}\\ \text{(file } B) \quad \widehat{Y}|x &\sim N_{pn_B}(\widehat{\mu}_{Y|ZX}, \widehat{\Sigma}_{Y|ZX} \otimes I_{nB}) \quad \text{with}\end{aligned} \tag{4.42}$$

$$\begin{aligned}\widehat{\mu}_{X|ZY} &= Z_A\widehat{\beta}_{XZ.Y} + Y\widehat{\beta}_{XY.Z} = Z_A\widehat{\beta}_{XZ} + (Y - Z_A\widehat{\beta}_{YZ})\widehat{\Sigma}^{-1}_{Y|Z}\widehat{\Sigma}_{21}\,,\\ \widehat{\mu}_{Y|ZX} &= Z_B\widehat{\beta}_{YZ.X} + X\widehat{\beta}_{YX.Z} = Z_B\widehat{\beta}_{YZ} + (X - Z_B\widehat{\beta}_{XZ})\widehat{\Sigma}^{-1}_{X|Z}\widehat{\Sigma}_{12}.\end{aligned}$$

In the frequentist regression imputation with random residual, called RIEPS hereinafter, the missing values are imputed according to (4.42). The parameters are estimated from the observed data as described above. In the next section we extend this approach to its Bayesian version, performing random draws for the parameters instead of estimating them.

4.4.3 Discussion

After having predicted the values according to their regression on all observed data (4.40) Rubin (1986) introduced predictive mean matching to impute the missing values. With predictive mean matching the observed x_i value of a unit i from file B is imputed to another unit j in file A, if their predicted values $\widehat{x}_{A,i}$ and $\widehat{x}_{B,j}$ have a minimal distance. The same is done with all missing values of Y. Predictive mean matching has the advantage over the conventional single regression imputation that only eligible values of the missing variable are imputed; for example, a binary variable is always assigned one of its two possible outcomes.[10] Some hot deck imputation procedures may be regarded as special cases of predictive mean matching; see Little (1988). Moreover, one may define a pool of nearest donor units to a particular unit with missing values and sample with or without replacement from this set to provide m imputations.

Regression imputation is a computationally interesting approach, because neither random draws for the parameters nor iterations to achieve any stationary distribution are necessary. Despite its computational advantages and its general acceptance among practitioners, this imputation procedure is not Bayesianly proper because the parameters are not randomly drawn according to their observed-data posterior. Also it is not proper in the sense defined by Rubin (1987). Roughly speaking, regression imputation often yields to biased estimates due to its lack of asymptotic properties.[11] We discuss and illustrate this feature in the simulation study at the end of this chapter.

4.5 Noniterative Multivariate Imputation Procedure

4.5.1 Introduction

Extending the regression imputation procedure already proposed we now introduce a new non-iterative Bayesian-based imputation procedure which is Bayesianly proper by definition. We call it **NIBAS** for short hereinafter. The elements of each column of the common matrix Z can be an appropriate sequence of -1s, 0s, or 1s corresponding to a design matrix mapping an experimental situation, contain 0s and 1s describing some qualitative variables, or, finally, may simply contain continuous values. However, we do not need to make any distributional assumptions about Z; we only require Z being suitable to serve as predictor matrix in a linear regression model.

Concerning the variables never jointly observed our assumptions are more restrictive. Here we require variables that may be regarded, at least, as being univariate normally distributed after some transformations have been done. Variables concerning media and consuming behavior may fulfill this demand after applying some useful transformations; see, e.g., O'Brien (1991) or Wendt (2000). In Chapter 5 we discuss situations where the actual data are surely not normal but the transformed data are more or less normal, at least enough for our practical purposes.

---

[10] However, imputed values are also often rounded off to the next feasible value; see Schafer and Olsen (1999).

[11] See, for example, the results of a simulation study performed by Rässler (2000).

4.5.2 Imputation Procedure for Univariate Variables

Again we assume the general linear model for both data sets according to

$$\begin{aligned} &(\text{file } A) \quad Y = Z_A\beta_{YZ} + U_A, \qquad U_A \sim N_{n_A}(0, \sigma_A^2 I_{n_A}), \\ &(\text{file } B) \quad X = Z_B\beta_{XZ} + U_B, \qquad U_B \sim N_{n_B}(0, \sigma_B^2 I_{n_B}), \end{aligned} \quad (4.43)$$

with Z_A and Z_B being disjoint parts of the common derivative matrix Z. The maximum likelihood estimates are again calculated via $\widehat{\beta}_{YZ} = (Z_A'Z_A)^{-1}Z_A'Y$ and $\widehat{\beta}_{XZ} = (Z_B'Z_B)^{-1}Z_B'X$. A suitable noninformative prior with β and $\log\sigma$ approximately independent and locally uniform is assumed with

$$f_{\beta_{YZ},\beta_{XZ},\sigma^2_{X|Z},\sigma^2_{Y|Z},\rho_{XY|Z}} = f_{\beta_{YZ},\beta_{XZ}} f_{\sigma^2_{X|Z},\sigma^2_{Y|Z}|\rho} f_\rho \propto \sigma_X^{-2}\sigma_Y^{-2} f_\rho, \quad (4.44)$$

and, hence, the marginal posterior distributions of $\sigma^2_{X|Z}|x$, $\sigma^2_{Y|Z}|y$ are inverted χ^2 distributions; i.e.,

$$\begin{aligned} \sigma^2_{Y|Z}|y &\sim \sum_{i=1}^{n_A}(y_i - Z_A\widehat{\beta}_{YZ})^2\chi^{-2}_{n_A-k} \quad \text{and} \\ \sigma^2_{X|Z}|x &\sim \sum_{i=1}^{n_B}(x_i - Z_B\widehat{\beta}_{XZ})^2\chi^{-2}_{n_B-k}. \end{aligned}$$

The conditional posterior distribution of β_{YZ} given $\sigma^2_{Y|Z}$ and the data is the multivariate normal distribution $N_k(\widehat{\beta}_{YZ}, \sigma^2_{Y|Z}(Z_A'Z_A)^{-1})$ and $N_k(\widehat{\beta}_{XZ}, \sigma^2_{X|Z}(Z_B'Z_B)^{-1})$; see Box and Tiao (1992), p. 115. We immediately find that the posterior distribution of $\rho_{XY|Z}$ equals its prior distribution; see Rubin (1987), p. 187. Thus, $\rho_{XY|Z}$ can be drawn from its prior or several levels of $\rho_{XY|Z}$ can be set arbitrarily to investigate sensitivity to $\rho_{XY|Z}$.

Having generated random values for the parameters from their posterior distributions we may draw values for the missing X of data set A from their regression on the observed Z and Y and for the missing Y of data set B from their regression on the observed Z and X, i.e., their conditional predictive distribution.

Thus we propose the following (at least Bayesianly proper) multiple imputation scheme allowing the use of continuous as well as discrete variables as common variables.

Algorithm (1)

- Compute the ordinary least square estimates $\widehat{\beta}_{XZ}$ and $\widehat{\beta}_{YZ}$ from the regression of each data set.
- Calculate each sum $(y - Z_A\widehat{\beta}_{YZ})'(y - Z_A\widehat{\beta}_{YZ})$ and $(x - Z_B\widehat{\beta}_{XZ})'(x - Z_B\widehat{\beta}_{XZ})$ of squared residual errors with $y = (y_1, y_2, \ldots, y_{n_A})'$, $x = (x_1, x_2, \ldots, x_{n_B})'$.

- Choose a value for $\rho_{XY|Z}$ either from its prior or several arbitrary levels.
- Perform random draws for the parameters from their observed-data posterior distribution according to the following routine.

$$
\begin{array}{lll}
\text{Step 1:} & \sigma^2_{Y|Z}|y & \sim & \frac{\sum_{i=1}^{n_A}(y_i - Z_A\widehat{\beta}_{YZ})^2}{\chi^2_{n_A-k}}, \\
& \sigma^2_{X|Z}|x & \sim & \frac{\sum_{i=1}^{n_B}(x_i - Z_B\widehat{\beta}_{XZ})^2}{\chi^2_{n_B-k}}. \\
\text{Step 2:} & \beta_{YZ}|\sigma^2_{Y|Z}, y & \sim & N_k(\widehat{\beta}_{YZ}, \sigma^2_{Y|Z}(Z'_A Z_A)^{-1}), \\
& \beta_{XZ}|\sigma^2_{X|Z}, x & \sim & N_k(\widehat{\beta}_{XZ}, \sigma^2_{X|Z}(Z'_B Z_B)^{-1}). \\
\text{Step 3:} & \text{Set } \sigma_{XY|Z} & = & \rho_{XY|Z}\sqrt{\sigma^2_{X|Z}\sigma^2_{Y|Z}}, \\
& \beta_{XY.Z} & = & \sigma_{XY|Z}/\sigma^2_{Y|Z}, \text{ and} \\
& \beta_{YX.Z} & = & \sigma_{XY|Z}/\sigma^2_{X|Z}. \\
\text{Step 4:} & X|y, \beta, \sigma^2, \rho & \sim & N_{n_A}(Z_A\beta_{XZ} + \beta_{XY.Z}(y - Z_A\beta_{YZ}); \\
& & & \sigma^2_{X|Z}(1 - \rho^2_{XY|Z})I_{n_A}) \\
& Y|x, \beta, \sigma^2, \rho & \sim & N_{n_B}(Z_B\beta_{YZ} + \beta_{YX.Z}(x - Z_B\beta_{XZ}); \\
& & & \sigma^2_{Y|Z}(1 - \rho^2_{XY|Z})I_{n_B})
\end{array}
$$

This imputation routine is very similar to the regression imputation method proposed in the section before; for instance, compare Step 4 of Algorithm (1) with (4.42). Yet here we perform random draws for the parameters instead of taking the estimates from the data. More uncertainty to account for the missing data is incorporated than by simply adding a random residual. Moreover, we are able to influence the resulting imputations by the prior choice of $\rho_{XY|Z}$.

When a small but complete sample is available, this sample may be used to estimate the conditional association of Y and X first instead of setting a prior value for $\rho_{XY|Z}$. Then the algorithm will continue as described above exploiting all the available information.

4.5.3 Imputation Procedure for Multivariate Variables

Making use of Theorem 8.5.1 by Box and Tiao (1992), p. 461, the extensions of Algorithm (1) to higher-dimensional Y and X are straightforward.

Again we assume the general linear model for both data sets with

$$
\begin{array}{lllll}
(\text{file } A) & Y & = & Z_A\beta_{YZ} + U_A, & U_A \sim N_{pn_A}(0, \Sigma_{22} \otimes I_{n_A}), \\
(\text{file } B) & X & = & Z_B\beta_{XZ} + U_B, & U_B \sim N_{qn_B}(0, \Sigma_{11} \otimes I_{n_B}),
\end{array} \quad (4.45)
$$

with Z_A and Z_B again denoting the corresponding parts of the common derivative matrix Z.

Again we assume a multivariate normal data model such as (4.33) for $(X, Y|Z = z) = (X_1, X_2, \ldots, X_q, Y_1, Y_2, \ldots, Y_p|Z = z)$ with expectation

$\mu_{XY|Z}$ and covariance matrix $\Sigma_{XY|Z}$; i.e.,

$$\Sigma_{XY|Z} = \left(\begin{array}{ccc|ccc} \sigma_{X_1X_1|Z} & \cdots & \sigma_{X_1X_q|Z} & \sigma_{X_1Y_1|Z} & \cdots & \sigma_{X_1Y_p|Z} \\ \cdots & \cdots & \cdots & \cdots & \cdots & \cdots \\ \sigma_{X_qX_1|Z} & \cdots & \sigma_{X_qX_q|Z} & \sigma_{X_qY_1|Z} & \cdots & \sigma_{X_qY_p|Z} \\ \hline \sigma_{Y_1X_1|Z} & \cdots & \sigma_{Y_1X_q|Z} & \sigma_{Y_1Y_1|Z} & \cdots & \sigma_{Y_1Y_p|Z} \\ \cdots & \cdots & \cdots & \cdots & \cdots & \cdots \\ \sigma_{Y_pX_1|Z} & \cdots & \sigma_{Y_pX_q|Z} & \sigma_{Y_pY_1|Z} & \cdots & \sigma_{Y_pY_p|Z} \end{array}\right) = \begin{pmatrix} \Sigma_{11} & \Sigma_{12} \\ \Sigma_{21} & \Sigma_{22} \end{pmatrix},$$

whereas $\Sigma_{11} = \Sigma_{X|Z}$, $\Sigma_{22} = \Sigma_{Y|Z}$ and

$$R_{XY|Z} = \begin{pmatrix} \rho_{X_1Y_1|Z} & \cdots & \rho_{X_1Y_p|Z} \\ \cdots & \cdots & \cdots \\ \rho_{X_qY_1|Z} & \cdots & \rho_{X_qY_p|Z} \end{pmatrix} = \{\rho_{X_iY_j|Z}\}$$

for $i = 1, 2, \ldots, q$, $j = 1, 2, \ldots, p$.

As a suitable noninformative prior we assume independence between β and Σ choosing

$$\begin{aligned} f_{\beta_{YZ},\beta_{XZ},\Sigma_{X|Z},\Sigma_{Y|Z},R_{XY|Z}} &= f_{\beta_{YZ},\beta_{XZ}} f_{\Sigma_{X|Z}|R_{XY|Z}} f_{\Sigma_{Y|Z}|R_{XY|Z}} f_{R_{XY|Z}} \\ &\propto \Sigma_{X|Z}^{-((q+1)/2)} \Sigma_{Y|Z}^{-((p+1)/2)} f_{R_{XY|Z}}. \end{aligned} \quad (4.46)$$

The joint posterior distribution for the matching case can be factored into the prior and likelihood derived by file A and file B, respectively. The same holds if μ is defined by the regression on Z; i.e., $\mu_X = Z_B\beta_{XZ}$ and $\mu_Y = Z_A\beta_{YZ}$. Then the joint posterior distribution can be factored with

$$\begin{aligned} & f_{\beta_{XZ},\beta_{YZ},\Sigma_{X|Z},\Sigma_{Y|Z},R_{XY|Z}|X,Y} = \\ & c_X^{-1} L(\beta_{XZ}, \Sigma_{X|Z}; x) f_{\Sigma_{X|Z}|R_{XY|Z}} \\ & c_Y^{-1} L(\beta_{YZ}, \Sigma_{Y|Z}; y) f_{\Sigma_{Y|Z}|R_{XY|Z}} f_{R_{XY|Z}}. \end{aligned} \quad (4.47)$$

Thus, our problem of specifying the posterior distributions reduces to standard derivation tasks described, for example, by Tiao and Zellner (1964) or Box and Tiao (1992), p. 439.

Algorithm (2)

- Compute the ordinary least squares estimates $\widehat{\beta}_{YZ} = (Z'_A Z_A)^{-1} Z'_A Y$ and $\widehat{\beta}_{XZ} = (Z'_B Z_B)^{-1} Z'_B X$ from the regression of each data set. Note that $\widehat{\beta}_{YZ}$ is a $k \times p$ matrix and $\widehat{\beta}_{XZ}$ is a $k \times q$ matrix of the OLS or ML estimates of the general linear model.
- Calculate the following matrices proportional to the sample covariances for each regression with

$$S_Y = (Y - Z_A\widehat{\beta}_{YZ})'(Y - Z_A\widehat{\beta}_{YZ}),$$

$$S_X = (X - Z_B\widehat{\beta}_{XZ})'(X - Z_B\widehat{\beta}_{XZ}).$$

- Choose a value for the correlation matrix $R_{XY|Z}$ or each $\rho_{X_iY_j|Z}$ for $i = 1, 2, \ldots, q, j = 1, 2, \ldots, p$
  - (a) from its prior according to some distributional assumptions, i.e., uniform over the $p + q$-dimensional]-1,1[-space, or
  - (b) several arbitrary levels, or
  - (c) estimate a value from a small but completely observed data set.

  The latter should be the most realistic case in many practical situations and is used among others for the evaluation study in Chapter 5.

- Perform random draws for the parameters from their observed data posterior distribution according to the following scheme.

$$\begin{array}{llcl}
\text{Step 1:} & \Sigma_{22}|y & \sim & W_p^{-1}(v_A, S_Y^{-1})\, v_A = n_A - (k+p) + 1 \\
 & \Sigma_{11}|x & \sim & W_q^{-1}(v_B, S_X^{-1})\, v_B = n_B - (k+q) + 1 \\
\text{Step 2:} & \beta_{YZ}|\Sigma_{22}, y & \sim & N_{pk}(\widehat{\beta}_{YZ}, \Sigma_{22} \otimes (Z_A'Z_A)^{-1}), \\
 & \beta_{XZ}|\Sigma_{11}, y & \sim & N_{qk}(\widehat{\beta}_{XZ}, \Sigma_{11} \otimes (Z_B'Z_B)^{-1}), \\
\text{Step 3:} & \text{Set } \Sigma_{12} & = & \{\sigma_{X_iY_j|Z}\} \text{ with } \sigma_{X_iY_j|Z} \\
 & & = & \rho_{X_iY_j|Z}\sqrt{\sigma^2_{X_i|Z}\sigma^2_{Y_j|Z}} \\
 & & & \text{with } \sigma^2_{X_i|Z}, \sigma^2_{Y_j|Z} \text{ derived by Step 1} \\
 & & & \text{for } i = 1, 2, \ldots, q, j = 1, 2, \ldots, p. \\
\text{Step 4:} & X|y, \beta, \Sigma & \sim & N_{qn_A}\left(Z_A\beta_{XZ} + (Y - Z_A\beta_{YZ})\Sigma_{22}^{-1}\Sigma_{21};\right. \\
 & & & \left.(\Sigma_{11} - \Sigma_{12}\Sigma_{22}^{-1}\Sigma_{21}) \otimes I_{n_A}\right) \\
 & Y|x, \beta, \Sigma & \sim & N_{pn_B}\left(Z_B\beta_{YZ} + (X - Z_B\beta_{XZ})\Sigma_{11}^{-1}\Sigma_{12};\right. \\
 & & & \left.(\Sigma_{22} - \Sigma_{21}\Sigma_{11}^{-1}\Sigma_{12}) \otimes I_{n_B}\right).
\end{array}$$

We achieve $\Sigma_{X|Z}|x$ and $\Sigma_{Y|Z}|y$ each following an inverted-Wishart distribution. The conditional posterior distribution of β_{XZ} (β_{YZ}) given $\Sigma_{X|Z}$ ($\Sigma_{Y|Z}$) and the observed data is a multivariate normal distribution. The posterior distribution of $R_{XY|Z}$ again equals its prior distribution. Having thus obtained the observed-data posteriors and the conditional predictive distributions a Bayesianly proper multiple imputation procedure for multivariate variables X and Y can be proposed with Algorithm (2).

The similarity of this new proposed procedure with the regression imputation method according to Section 4.4 and especially with (4.42) is obvious. It should be mentioned again that conditioning on $Z = z$ in this procedure is not the correct formulation because we do not make any distributional assumption about Z. Thus, the predictive power of the common variables Z especially affects the last step. Choosing the correlation matrix $R_{XY|Z}$ uniform over the $p + q$-dimensional]-1,1[-space may lead to invalid conditional variances in Step 4. To achieve imputations reflecting the bounds we propose to set $R_{XY|Z} = 0_{q \times p}$ and add some $\pm\epsilon$ iteratively until the variance matrices in Step 4 are no longer positive definite. A similar procedure to get admissible values of the covariance matrix is proposed by Moriarity and Scheuren (2001) for the multivariate case.[12]

Notice that any r-dimensional Wishart distributed random variable with parameters Σ and v can be generated easily (see Fahrmeir et al. (1996), p. 38) as follows.

- Compute v data vectors of $X = (X_1, \dots, X_r) \sim N_r(0, \Sigma)$,
- then, $W = X'X = \sum_{i=1}^{v} x_i x_i' \sim W_r(v, \Sigma)$ is Wishart distributed with v degrees of freedom.

In the case of univariate X and Y, Algorithm (1) is obtained therefrom, since $W_r(v, a) = a\chi^2(v)$ if $r = 1$.

The noniterative multivariate imputation procedure as proposed above requires $\frac{1}{2}(p+q)(p+q+1)$ distinct variances and covariances as well as $k(p+q)$ regression coefficients. Thus, the number of parameters operated by NIBAS is kept modest with

$$\frac{1}{2}(p+q)(p+q+1) + k(p+q), \tag{4.48}$$

which is due to the fact that the common variables are only used as independent variables in the regression instead of being modelled themselves. This could be an interesting feature, especially when the number of variables is large relative to the number of subjects, which is typical for matching tasks.

4.5.4 Discussion

We have shown that it is possible for the matching problem to formulate a suitable data model and prior distribution and derive the observed-data posterior therefrom. This is due to the special missingness pattern induced by the matching situation. The method proposed here is Bayesianly proper by definition. Remember that we are not able to estimate the association of the variables never jointly observed from the data without using prior information. Thus, we may never verify whether $\widehat{E}(\widehat{\theta}_{MI}) \approx \widehat{\theta}$ and $\widehat{var}(W) \approx \widehat{var}(\widehat{\theta})$ for the unobserved association, which is requested to achieve a frequentist proper imputation procedure.

Nevertheless, this approach is rather encouraging because it allows a quick and controlled statistical matching generating multiple imputations which are at least Bayesianly proper. Moreover, the number of variables and observations is limited only by the memory and storage of S-PLUS software since the imputation routine is noniterative. Further criteria and

---

[12] For univariate X and Y variables $\rho_{XY|Z}$ can be chosen from $]-1,1[$ using Algorithm (1). The bounds can be calculated directly then; see also Moriarity and Scheuren (2001).

advantages of NIBAS are discussed in the simulation study at this chapter's end as well as in Chapter 5.

In conventional missing data settings more often Markov chain Monte Carlo techniques are applied based on the usually easier to operate complete-data posterior distributions. The application of these techniques to the matching task is discussed in the following section.

4.6 Data Augmentation

4.6.1 Introduction

A Markov chain Monte Carlo (MCMC) method ideally suited to missing data problems in general is the data augmentation algorithm of Tanner and Wong (1987), which was primarily developed to simulate the posterior distribution of any parameters of interest by means of Markov chains.

MCMC is a collective term for methods that generate random numbers which can be thought of as realizations from a desired probability distribution through Markov chains. A Markov chain is a sequence $\{U^{(t)} : t = 0, 1, \ldots\}$ of any random variables $U^{(t)}$, which distribution only depends on the previous one $U^{(t-1)}$, i.e.,

$$\{U^{(t)} : t = 0, 1, \ldots\} \quad \text{with} \quad f_{U^{(t)}|U^{(0)},\ldots,U^{(t-1)}} = f_{U^{(t)}|U^{(t-1)}}.$$

Markov chains are thus constructed from the so-called transition kernel $f_{U^{(t)}|U^{(t-1)}}$. For details about the construction of MCMC methods see, e.g., Robert and Casella (1999). When proceeding along the sequence, under certain regularity conditions, the distributions of the elements stabilize to the stationary distributions f_U of interest, becoming independent from any arbitrary starting point. Repeated simulations of such a chain allow us to simulate draws from the desired distribution. A very popular method of MCMC is the Gibbs sampler. The data augmentation method corresponds to the particular case of a Gibbs sampler, when two conditional distributions $f_{U_1|U_2}$ and $f_{U_2|U_1}$ are available. Note that both U_1 and U_2 can denote either scalars or vectors. A comprehensive overview of data augmentation algorithms and their applications are given by van Dyk and Meng (2001).

The application of data augmentation in the context of missing data is principally as simple as appealing. Instead of calculating the often difficult to derive posterior predictive distributions of the missing data or the observed-data posterior distributions of the parameters, we realize an iterative two-step process. First, the missing values are sampled from their conditional predictive distribution $U_{mis}|U_{obs}, \Theta \sim f_{U_{mis}|U_{obs},\Theta}(u_{mis}|u_{obs}, \theta^{(t)})$ given an actual draw of the parameter $\theta^{(t)}$. Second, the unknown parameters are sampled from the simulated complete-data posterior $\Theta|U_{obs}, U_{mis} \sim f_{\Theta|U_{obs},U_{mis}}(\theta|u_{obs}, u_{mis}^{(t)})$ given the actually imputed values $u_{mis}^{(t)}$. Given an initial value this defines a Markov chain

$\{(U_{mis}^{(t)}, \Theta^{(t)}) : t = 0, 1, \ldots\}$, which, under quite general conditions, converges to the stationary distribution $f_{U_{mis},\Theta|U_{obs}}(u_{mis}, \theta|u_{obs})$. If we proceed along the sequences a large number of times, we expect to produce draws of Θ from its observed-data posterior $f_{\Theta|U_{obs}}$ and draws of U_{mis} from its posterior predictive distribution $f_{U_{mis}|U_{obs}}$. Thus, finally, we have achieved Bayesianly proper imputations for U_{mis} from its posterior predictive distribution $f_{U_{mis}|U_{obs}}$ according to (4.20).

Following Rubin et al. (1995) the data augmentation algorithm can be viewed as the Bayesianly stochastic counterpart of the EM algorithm. As with the latter, data augmentation is performed via two steps: the imputation-step (I-step) and the posterior-step (P-step).

The I-step generates random values for the missing data U_{mis} according to their conditional predictive distribution $f_{U_{mis}|U_{obs},\theta}$ given the observed data and an actual draw of the parameter θ; i.e.,

$$U_{mis}^{(t+1)}|u_{obs}, \theta^{(t)} \sim f_{U_{mis}|U_{obs},\Theta}(u_{mis}|u_{obs}; \theta^{(t)}). \tag{4.49}$$

This is the stochastic version of the E-step, imputing a random draw for U_{mis} instead of its expectation.

The P-step draws new actual values for θ according to their complete-data posterior distribution $f_{\Theta|U_{obs},U_{mis}}$ given the observed data and actual draws of the missing data; i.e.,

$$\Theta^{(t+1)}|u_{obs}, u_{mis}^{(t+1)} \sim f_{\Theta|U_{obs},U_{mis}}(\theta|u_{obs}, u_{mis}^{(t+1)}). \tag{4.50}$$

Such a random draw of Θ is the Bayesianly stochastic counterpart of maximizing the complete-data likelihood being performed in the M-step. Analogous to the EM, which uses the complete-data likelihood, data augmentation makes use of the complete-data posterior, which often is more attractive than the observed-data posterior.

Finally, repeating this procedure many times yields a Markov chain $\{(U_{mis}^{(t)}, \Theta^{(t)}) : t = 0, 1, \ldots\}$ with stationary distribution $f_{U_{mis},\Theta|U_{obs}}$ as well as $\{\Theta^{(t)} : t = 0, 1, \ldots\}$ with stationary distribution $f_{\Theta|Y_{obs}}$ and $\{U_{mis}^{(t)} : t = 0, 1, \ldots\}$ with stationary distribution $f_{U_{mis}|U_{obs}}$. Thus, data augmentation provides rather flexible tools for creating multiple imputations from parametric models.

4.6.2 Imputation Procedure

We do not want to present all the different data augmentation algorithms here, since they are described and derived in detail by Schafer (1997). For our practical purposes we treat the normal data model; actually this is the most flexible one concerning the number of variables being incorporated in any realistic matching task. Moreover, a standalone MS-Windows^{TM} version called NORM is available for free; see Schafer (1999b).

NORM uses a data model quite similar to NIBAS but the parametric data model is applied to the complete data whereas NIBAS makes use of the observed-data posterior. Since the complete-data posterior does not take into account the pattern of the missing data, data augmentation is usually applied to a wide range of missing data problems. In particular, data augmentation has not yet been applied explicitly to the matching case because then the identification problems inherent in statistical matching have to be surmounted. Therefore we give a detailed description of how data augmentation may be used in the context of statistical matching.

The complete data model $f_{U|\mu,\Sigma}$ for $(X, Y, Z) = U = (U_{mis}, U_{obs})$ used by NORM is

$$U|\mu, \Sigma \sim N_r(\mu, \Sigma), \quad r = k + q + p, \quad \begin{pmatrix} \Sigma_{XX} & \Sigma_{XY} & \Sigma_{XZ} \\ \Sigma_{YX} & \Sigma_{YY} & \Sigma_{YZ} \\ \Sigma_{ZX} & \Sigma_{ZY} & \Sigma_{ZZ} \end{pmatrix}.$$

The predictive distributions $f_{U_{mis}|U_{obs},\mu,\Sigma}$ of missing variables X and Y are given by

$$\begin{aligned} (\text{file } A) \quad X|z, y, \mu, \Sigma &\sim N_q(\mu_{X|ZY}, \Sigma_{X|ZY}), \\ (\text{file } B) \quad Y|z, x, \mu, \Sigma &\sim N_p(\mu_{Y|ZX}, \Sigma_{Y|ZX}), \end{aligned}$$

for each unit of files A and B, respectively. Notice that these are the multivariate predictive distributions; their univariate versions were already introduced with the EM algorithm in (4.7).

For clarity we distinguish among several cases as suitable prior distributions and thus as posterior distributions (the derivations are given in the appendix).

(1) Assume μ and Σ approximately independent and μ uniform over the r-dimensional space. Take a noninformative prior with $f_\Sigma \propto |\Sigma|^{-(r+1)/2}$. We get the posterior distribution with

$$\begin{aligned} \Sigma|u &\sim W_r^{-1}\left(n - r, S(\overline{u})^{-1}\right) \text{ and} \\ \mu|\Sigma, u &\sim N_r\left(\overline{u}, \frac{1}{n}\Sigma\right); \end{aligned}$$

see Schafer (1997), p. 154. Remember that $S(\overline{u})$ is proportional to the sample covariance matrix; see also the appendix.

(2) Furthermore, let us apply a noninformative prior for μ again and an informative prior for Σ with $f_\Sigma \propto |\Sigma|^{-(\frac{1}{2}\nu+r)} \exp\left\{-\frac{1}{2}tr(\Sigma^{-1}\Lambda^{-1})\right\}$. Then we get

$$\begin{aligned} \Sigma|u &\sim W_r^{-1}(\nu + n - 1, \Lambda'), \quad \Lambda' = (\Lambda^{-1} + S(\overline{u}))^{-1} \text{ and} \\ \mu|\Sigma, u &\sim N_r\left(\overline{u}, \frac{1}{n}\Sigma\right) \end{aligned}$$

for some hyperparameters ν and Λ.

(3) Finally we may apply a so-called ridge prior, as proposed by Schafer (1997), p. 156, i.e., assuming

$$f_{\mu,\Sigma} \propto |\Sigma|^{-(\nu/2+r)-1/2} \exp\left\{-\frac{1}{2}tr(\Sigma^{-1}\Lambda^{-1})\right\},$$

which can be shown to be some limiting form of the inverted-Wishart distribution. Thus we derive the posterior distribution according to

$$\begin{aligned}\Sigma|u &\sim W_r^{-1}(\nu+r+n-1,\Lambda'), \quad \Lambda' = (\Lambda^{-1}+S(\overline{u}))^{-1} \text{ and}\\ \mu|\Sigma,u &\sim N_r\left(\overline{u},\frac{1}{n}\Sigma\right)\end{aligned}$$

for some hyperparameters ν and Λ; see Schafer (1997), p. 152.[13]

The default prior implemented in NORM is the latter. For our point of interest it does not matter which informative prior is used; thus we may use the ready-to-operate version NORM for fusing the data. Having specified the prior the following steps are performed iteratively by NORM.

IMPUTATION-STEP: For each unit i of file A or file B, respectively, a random draw is performed independently, because we assumed the data to be independent realizations from the underlying population. Hence the missing values of X and Y are drawn according to their conditional predictive distribution

$$\begin{aligned}\text{(file A)}\ X^{(t+1)}|z,y,\mu^{(t)},\Sigma^{(t)} &\sim N_{qn_A}\left(\mu^{(t)}_{X|ZY},\Sigma^{(t)}_{X|ZY}\otimes I_{n_A}\right),\\ \text{(file B)}\ Y^{(t+1)}|z,x,\mu^{(t)},\Sigma^{(t)} &\sim N_{pn_B}\left(\mu^{(t)}_{Y|ZX},\Sigma^{(t)}_{Y|ZX}\otimes I_{n_B}\right),\end{aligned} \tag{4.51}$$

given the observed data and an actual draw of the parameters $\mu^{(t)}$ and $\Sigma^{(t)}$. The actual values of the conditional expectation and variances are calculated from $\mu^{(t)}$ and $\Sigma^{(t)}$ and the completed data as follows.

$$\begin{aligned}\mu_{X|ZY} &= \mu_X + (Z-\mu_Z \quad Y-\mu_Y)\Sigma_{11}^{-1}\Sigma_{12},\\ \Sigma_{X|ZY} &= \Sigma_{XX} - \Sigma_{12}'\Sigma_{11}^{-1}\Sigma_{12} \text{ where}\\ \Sigma_{11} &= \begin{pmatrix}\Sigma_{ZZ} & \Sigma_{ZY}\\ \Sigma_{ZY} & \Sigma_{YY}\end{pmatrix}, \quad \Sigma_{12} = \begin{pmatrix}\Sigma_{ZX}\\ \Sigma_{XY}\end{pmatrix}.\\ \mu_{Y|ZX} &= \mu_Y + (Z-\mu_Z \quad X-\mu_X)\Sigma_{11}^{-1}\Sigma_{12},\\ \Sigma_{Y|ZX} &= \Sigma_{YY} - \Sigma_{12}'\Sigma_{11}^{-1}\Sigma_{12}, \text{ where}\end{aligned}$$

[13]Notice that Schafer (1997) takes parameters similar to Anderson (1984), p. 268, denoting the Wishart distribution with m degrees of freedom. Our notation follows Box and Tiao (1992), so that $m = v + r - 1$.

$$\Sigma_{11} = \begin{pmatrix} \Sigma_{ZZ} & \Sigma_{ZX} \\ \Sigma_{ZX} & \Sigma_{XX} \end{pmatrix}, \quad \Sigma_{12} = \begin{pmatrix} \Sigma_{ZY} \\ \Sigma_{XY} \end{pmatrix}.$$

POSTERIOR-STEP: Given the drawn values of $\widehat{U}_{mis} = X^{(t+1)}$ in file A and $\widehat{U}_{mis} = Y^{(t+1)}$ in file B, draw a new value for μ and Σ from its complete posterior distribution

$$\begin{aligned}
(1)\ \Sigma^{(t+1)}|\widehat{u}_{mis}, u_{obs} &\sim W_r^{-1}(n-r, S(\overline{u})^{-1}), \text{ or} && (4.52)\\
(2)\ \Sigma^{(t+1)}|\widehat{u}_{mis}, u_{obs} &\sim W_r^{-1}(\nu+n-1, (\Lambda^{-1}+S(\overline{u}))^{-1}), \text{ or} && (4.53)\\
(3)\ \Sigma^{(t+1)}|\widehat{u}_{mis}, u_{obs} &\sim W_r^{-1}(\nu+r+n-1, (\Lambda^{-1}+S(\overline{u}))^{-1}), && (4.54)\\
\mu^{(t+1)}|\Sigma^{(t+1)}, \widehat{u}_{mis}, u_{obs} &\sim N_r(\overline{u}, n^{-1}\Sigma^{(t+1)}), && (4.55)
\end{aligned}$$

with $S(\overline{u})$ and $\overline{u}$ calculated from the completed data $U = (U_{obs}, \widehat{U}_{mis})$.

Thus we have realized a Gibbs sampling procedure by iterative random draws for $\widehat{U}_{mis}, \widehat{\Theta}$ from (I-step) the conditional predictive distribution of $U_{mis}|u_{obs}, \widehat{\theta}$, and (P-step) the complete-data posterior distribution of $\Theta|\widehat{u}_{mis}, u_{obs}$, to achieve a Markov chain with the desired stationary distributions of $f_{\Theta|U_{obs}}$ and $f_{U_{mis}|U_{obs}}$. Throughout the whole procedure $\frac{1}{2}r(r+1)$ distinct variances and covariances as well as r means are required, thus a number of

$$\frac{1}{2}r(r+1) + r = \frac{1}{2}(k+p+q)(k+p+q+1) + (k+p+q) \qquad (4.56)$$

parameters are to be operated, more than we use for NIBAS according to (4.48).

To create the final m multiple imputations, we may either perform m independent MCMC runs each with large t and use the latest t imputation as a solution or take every $(t+t^\star)$ imputation to get m multiple imputations with, e.g., $t^\star = 100$.

4.6.3 Discussion

A so-called "burn-in period" should be taken into account to achieve independence of the chain from the starting values. The interested reader is referred to discussions by Schafer (1997) about its length and the asymptotic behavior of the data augmentation algorithm in general.

When an uninformative prior is applied, the posteriors are based only on updated parameters estimated from the observed and imputed data. Since the covariance of the variables not jointly observed cannot be estimated from the data this may cause some problems. As we have seen before in (4.4), the observed-data likelihood is not influenced by the covariance of the two variables never jointly observed. Running the EM with different starting values will always yield different stationary values for Σ_{XY}. If we evaluate the observed-data likelihood at each stationary value, we easily see that it is exactly the same in each case; thus these values form a ridge of constant likelihood. Although the I- and the P-step are both well defined based on the complete-data posteriors the algorithm cannot converge to any stationary distribution.[14] Using the ridge prior provided by Schafer (1999b) or any other informative prior overcomes this problem of inestimability.

4.7 Iterative Univariate Imputations by Chained Equations

4.7.1 Introduction

It is said that iterative univariate imputations were first implemented by Kennickell (1991) and Kennickell and McManus (1994); see Schafer and Olsen (1999).[15] The intuitively appealing idea behind the iterative univariate imputation procedure is to overcome the problem of suitably proposing and fitting a multivariate model for mixtures of categorical and continuous data by reducing the multivariate imputation task to conventional regression models iteratively completed. In many surveys and especially in the matching task it may be difficult to propose a sensible joint distribution for all variables of interest. On the other hand there is a variety of procedures available for regression modeling of continuous and categorical univariate response variables such as ordered or unordered logit/probit models (see Greene (1997)). Thus any plausible regression model $f_{U_{mis}|U_{obs},\Theta}$ may be specified for predicting each univariate variable U_{mis} that has to be imputed given all the other variables. This approach is also known as regression switching, chained equations, or variable-by-variable Gibbs sampling; see van Buuren and Oudshoorn (1999). In the variable-by-variable Gibbs sampling approach it is also possible to include only relevant predictor variables, thus reducing the number of parameters.

4.7.2 Imputation Procedure

Basically, multivariate linear regression models with univariate response variables for the complete data are proposed. Let denote $U_i = X_i$ and

$$U_{(-i)} = (Z, X_1, \ldots, X_{i-1}, X_{i+1}, \ldots, X_q, Y)$$

for variables X_i, $i = 1, 2, \ldots, q$, and $U_i = Y_i$ with

$$U_{(-i)} = (Z, X, Y_1, \ldots, Y_{i-1}, Y_{i+1}, \ldots, Y_p)$$

[14]For a detailed discussion of the inestimability of parameters see Schafer (1997), pp. 204–206.

[15]Ready to use and available for free via the Internet is software called MICE which is a recent implementation of some iterative univariate imputation methods in S-PLUS; see van Buuren and Oudshoorn (2000).

for variables Y_i, $i = 1, 2, \ldots, p$, respectively. Then the usual multivariate linear regression

$$U_i = U_{(-i)}\beta_i + V_i, \quad V_i \sim N(0, \sigma_i^2 I_n) \tag{4.57}$$

is assumed for variables X_i, $i = 1, 2, \ldots, q$, as well as for variables Y_i, $i = 1, 2, \ldots, p$. For simplicity we map the constant of each regression to z_1, thus assuming k common variables Z including the constant. Notice that $U_{(-i)}$ may also be reduced to any subset of all possible predictor variables. The least squares estimates are $\widehat{\beta}_i = (U'_{(-i)}U_{(-i)})^{-1}U_{(-i)}U_i$ and $\widehat{\sigma}_i^2 = (u_i - U_{(-i)}\widehat{\beta})'(u_i - U_{(-i)}\widehat{\beta})/(n_i - k_c)$ with $n_i \in \{n_A, n_B\}$ for $i = 1, 2, \ldots, q$, or $i = 1, 2, \ldots, p$, and $k_c = k + p + q - 1$, if all available variables are incorporated in the regression model.

Combined with the conventional linear regression model (4.57) an improper prior f_{β,σ^2} for β_i and σ_i^2 of every regression may be assumed with

$$f_{\beta,\sigma^2}(\beta_i, \sigma_i^2) = f_\beta(\beta_i) f_{\sigma^2}(\sigma_i^2) \propto 1/\sigma_i^2\,, \tag{4.58}$$

for variables $U_i = X_i$, $i = 1, 2, \ldots, q$, as well as for variables $U_i = Y_i$, $i = 1, 2, \ldots, p$.

Analogous to Algorithm (1) (or see Rubin (1987), pp. 166–167), the conditional posterior distributions $f_{\beta_i,\sigma_i^2|U_{(-i)}}$ derived from (4.58) and (4.57) are given by

$$\begin{aligned} \sigma_i^2 | u_{(-i)} &\sim (u_i - U_{(-i)}\widehat{\beta}_i)'(u_i - U_{(-i)}\widehat{\beta}_i)\chi^{-2}_{n_i - k_c} \\ \beta_i | \sigma_i^2, u_{(-i)} &\sim N_{k_c}(\widehat{\beta}_i, \sigma_i^2 (U'_{(-i)}U_{(-i)})^{-1}), \end{aligned}$$

$n_i \in \{n_A, n_B\}$, for variables $U_i = X_i$, $i = 1, 2, \ldots, q$, as well as for variables $U_i = Y_i$, $i = 1, 2, \ldots, p$.

Thus, multiple imputations may be performed according to the following routine.

$$\begin{array}{lll} \text{Step 1:} & \sigma_i^2 | u_{(-i)} & \sim (u_i - U_{(-i)}\widehat{\beta}_i)'(u_i - U_{(-i)}\widehat{\beta}_i)\chi^{-2}_{n_i - k_c} \\ \text{Step 2:} & \beta_i | \sigma_i^2, u_{(-i)} & \sim N_{k_c}(\widehat{\beta}_i, \sigma_i^2 (U'_{(-i)}U_{(-i)})^{-1}) \\ \text{Step 3:} & U_{mis,i} | \beta_i, \sigma_i^2, u_{(-i)} & \sim N_{n_i}\left(U_{(-i)}\beta_i, \sigma_i^2 I_{n_i}\right) \end{array}$$

for variables $U_{mis,i} = X_i$ in file A, $i = 1, 2, \ldots, q$, as well as for variables $U_{mis,i} = Y_i$ in file B, $i = 1, 2, \ldots, p$.

It should be noted carefully that in the regression model any variables with missing values may also serve as predictor variables; i.e., $U_{(-i)}$ itself may contain missing values and therefore Steps 1 to 3 cannot be carried out unless these values have been imputed. This suggests an iterative imputation algorithm analogous to the Gibbs sampler. To start with, we may fill in all the missing values by a simple imputation method, for example, any nearest neighbor algorithm only based on the common variables. Then we perform Steps 1 to 3 for each variable $U_{mis,i}$ for $i = 1, 2, \ldots, q$ in file A and $i = 1, 2, \ldots, p$ in file B, using the most recent imputed data for $U_{(-i)}$

at each step. If we repeat this cycle over and over, the system may stabilize to a set of joint relationships among U_{mis}.

For categorical response variables logistic or polytomous regression imputation models may be applied. Using the S-PLUS library MICE a wide variety of such imputation routines may be implemented.[16]

Although the problem of dimensionality is reduced by the chained equations to simpler regression models the number of parameters that are to be operated is usually rather high. If linear regression models are used based on all available variables, the model requires $p+q$ variances and $k_c(p+q)$ regression coefficients; i.e.,

$$r(p+q) = (k+p+q)(p+q) \tag{4.59}$$

parameters.

4.7.3 Discussion

Although this procedure is very attractive because it reduces the multivariate modeling task to simpler regression models the statistical properties of the method are not well understood yet. As long as we only propose models for univariate response variables the joint distribution of U_{mis} may not actually exist. A "real" Gibbs sampler starts with an existing but intractable joint distribution for the variables of interest, iteratively generating random variables from easier to operate conditional distributions derived from its joint distribution. In the context of iterative univariate imputations the conditional distributions are specified in the hope that these conditional distributions will define a suitable joint model. Casella and George (1992) construct examples of conditional distributions that do not define any possible joint distribution. Schafer and Olsen (1999) provide an extensive discussion of iterative univariate imputations, stating that "in such cases, iteratively sampling from these 'conditionals' creates a Markov chain that never actually converges; the probability law describing the state of the system does not stabilize but continues to change from cycle to cycle. The practical implications of this phenomenon in iterative univariate imputation are unknown."

Although the subject of incompatible conditionals is still an open research problem, van Buuren and Oudshorn (2000) state that they and Brand (1999) have studied the performance of a variety of iterative univariate imputation routines based on possibly incompatible conditionals by means of simulation studies with encouraging results.

---

[16]For a number of possible regression models and their implementation as iterative univariate imputation routines see the extensive description by Brand (1999), pp. 93–97. Principally they are based on explicit imputation models proposed by Rubin (1987), pp. 166–170.

As long as we do not ensure the existence of a joint distribution the procedure is neither proper in Rubin's sense nor is it Bayesianly proper as defined by Schafer (1997). Iterative univariate imputations should at least suffice the latter definition, if the joint distribution of U_{mis} exists and the Markov chain is run long enough to ensure independence of the starting values.

4.8 Simulation Study – Multivariate Normal Data

The simulation study described hereinafter has several objectives. The first objective is to explore the efficiency of estimating the unconditional association of the variables never jointly observed based on the imputed data set when different prior information about their conditional association is used. The second objective is to investigate to what extent a third data source of complete nature may improve the estimation of the unconditional association. Finally, we demonstrate the application of the proposed matching techniques and highlight their benefits compared to the traditional approaches presented in the simulation study at the end of Chapter 2.

4.8.1 Data Model

Let (Z_1, Z_2, X, Y) each be univariate standard normally distributed variables with their joint distribution $(Z_1, Z_2, X, Y) \sim N_4(0, \Sigma)$ and let

$$\Sigma = \left(\begin{array}{cc|c|c} 1.0 & 0.2 & 0.5 & 0.8 \\ 0.2 & 1.0 & 0.5 & 0.6 \\ \hline 0.5 & 0.5 & 1.0 & \sigma_{XY} \\ \hline 0.8 & 0.6 & \sigma_{YX} & 1.0 \end{array}\right) = \left(\begin{array}{ccc} \Sigma_{ZZ} & \Sigma_{ZX} & \Sigma_{ZY} \\ \Sigma_{XZ} & \sigma_{XX} & \sigma_{XY} \\ \Sigma_{YZ} & \sigma_{YX} & \sigma_{YY} \end{array}\right),$$

$$\text{with } \Sigma_{XY} = \left(\begin{array}{cc} \sigma_{XX} & \sigma_{XY} \\ \sigma_{YX} & \sigma_{YY} \end{array}\right).$$

Throughout the study we assume that the true covariance is given with $\sigma_{XY} = \sigma_{YX} = 0.8$. Furthermore, let file $A = (Z_1, Z_2, Y)$ and file $B = (Z_1, Z_2, X)$; thus X and Y are never jointly observed.

As shown in Chapter 2, the nearest neighbor match leads to conditional independence of $X, Y|Z = z$ with the unconditional covariance after statistical matching $\widetilde{\sigma}_{XY} = \Sigma_{XZ}\Sigma_{ZZ}^{-1}\Sigma_{ZY} = 0.5833$. This value of $\widetilde{\sigma}_{XY} = 0.5833$ maximizes the determinant $|\Sigma|$ of the covariance matrix[17] as given above with all other parameters held fixed; thus we establish the maximal variation according to Wilks' generalized variance among Z, X, Y with a nearest neighbor match, as Figure 4.6 shows.

[17]This is also known as Wilks' generalized variance; see Mood et al. (1974), p. 353.

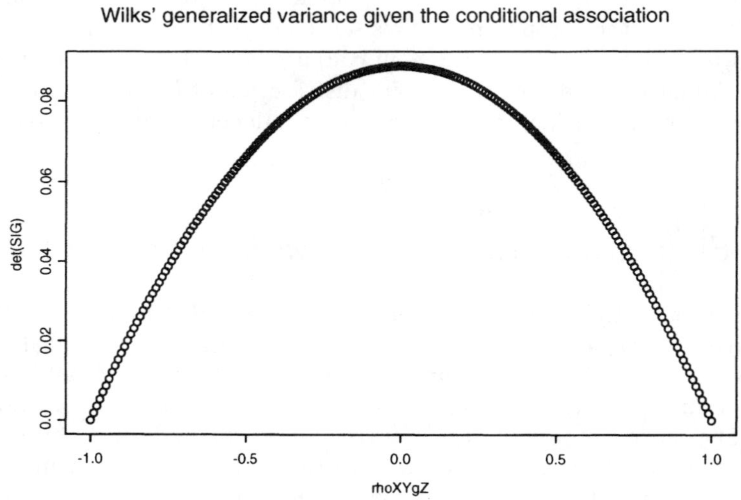

Figure 4.6. Determinant of Σ as function of $\rho_{XY|Z}$

To calculate the unconditional covariance, when a particular conditional correlation is given, the following well-known formulas are used.

$$\begin{aligned}\sigma_{XX|Z} &= \sigma_{XX} - \Sigma_{XZ}\Sigma_{ZZ}^{-1}\Sigma_{ZX},\\ \sigma_{YY|Z} &= \sigma_{YY} - \Sigma_{YZ}\Sigma_{ZZ}^{-1}\Sigma_{ZY},\\ \sigma_{XY|Z} &= \rho_{XY|Z}\sqrt{\sigma_{XX|Z}\sigma_{YY|Z}},\\ \sigma_{XY} &= \sigma_{XY|Z} + \Sigma_{XZ}\Sigma_{ZZ}^{-1}\Sigma_{ZY}.\end{aligned}$$

We also may directly compute

$$\begin{aligned}\Sigma_{XY} &= \Sigma_{XY|Z} + \begin{bmatrix}\Sigma_{XZ}\\ \Sigma_{YZ}\end{bmatrix}\Sigma_{ZZ}^{-1}\left[\Sigma_{ZX}\Sigma_{ZY}\right] \text{ with} \qquad (4.60)\\ \Sigma_{XY|Z} &= \begin{pmatrix}\sigma_{XX|Z} & \sigma_{XY|Z}\\ \sigma_{YX|Z} & \sigma_{YY|Z}\end{pmatrix}.\end{aligned}$$

According to the Cauchy–Schwarz inequality (see Mood et al. (1974), p. 162) the conditional covariance is bounded by $|\sigma_{XY|Z}| \leq \sqrt{\sigma_{XX|Z}\sigma_{YY|Z}} = \sqrt{0.5833 \cdot 0.1583} = 0.3039$. Thus we calculate some associations according to (4.60) with $\sigma_{XY} = \rho_{XY|Z} \cdot 0.3039 + 0.5833$ and the corresponding determinant values $|\Sigma|$ as displayed in Figure 4.6 and listed in Table 4.5.

In our setting is $\rho_{XY} = \sigma_{XY}$, because $\sigma_{XX} = \sigma_{YY} = 1$ and hence, only a range of the unconditional correlation of the two variables never jointly observed with $\rho_{XY} \in [0.2794 \quad 0.8872]$ yields a nonnegative definite matrix Σ.

$\rho_{XY\|Z}$	$\sigma_{XY\|Z}$	σ_{XY}	$\|\Sigma\|$
-1.0	-0.3039	0.2794	0.0000
-0.8	-0.2431	0.3402	0.0319
-0.6	-0.1823	0.4010	0.0567
-0.4	-0.1216	0.4618	0.0745
-0.2	-0.0608	0.5226	0.0851
0.0	0.0000	0.5833	0.0887
0.2	0.0608	0.6441	0.0851
0.4	0.1216	0.7049	0.0745
0.6	0.1823	0.7657	0.0567
0.8	0.2431	0.8265	0.0319
1.0	0.3039	0.8872	0.0000

Table 4.5. Values of the determinant $|\Sigma|$ as function of $\rho_{XY|Z}$

Based on this range we propose a new measure to display the explanatory power the common variables have and, thus, the validity a matching technique may achieve in a given context. Let $\rho^+_{X_iY_j}$ be the values of the positive maximum unconditional correlation of any two variables X_i, $i = 1, 2, \ldots, q$ and Y_j, $j = 1, 2, \ldots, p$ and $\rho^-_{X_iY_j}$ be the values of the negative maximum unconditional correlation. Then a suitable measure of the explanatory power of the common variables is given by the mean absolute difference

$$R_{AC} = \frac{1}{2qp} \sum_{i=1}^{q} \sum_{j=1}^{p} |\rho^+_{X_iY_j} - \rho^-_{X_iY_j}| \in [0, 1]. \tag{4.61}$$

In our example we find $\rho^+_{X_iY_j} = 0.8872$ and $\rho^-_{X_iY_j} = 0.2794$, thus, $R_{AC} = 0.3039$. In high-dimensional data structures R_{AC} may be estimated by means of the multiple imputation estimates $\widehat{\rho^+_{X_iY_j}}$ and $\widehat{\rho^-_{X_iY_j}}$ derived from the multiply imputed statistically matched files.

4.8.2 Design of the Study

We draw $n = 5000$ random numbers for (z, x, y) according to $(Z, X, Y) \sim N_4(0, \Sigma)$ as given with (2.18). This generated data set is divided into two parts, each file of size 2500 and all x (file A) or y (file B) values are eliminated. Then we either assume different conditional correlations of X and Y given $Z = z$ as prior information or take another random draw to generate a small but complete data source.

The multiple imputation procedures implemented for this study are

- NORM, the data augmentation algorithm assuming the normal model as described in Section 4.6 is applied based on $m = 5$ multiple imputations.[18] Using NORM we run a burn-in period of 100 iterations, then impute the missing data from every additional 50th iteration.

- NIBAS, the noniterative multivariate Bayesian regression model applying Algorithm (1) (see Section 4.5) based on $m = 5$ multiple imputations is applied.
- An iterative univariate imputation method according to Section 4.7 implemented via MICE is applied only to the conditional independence case, because it does not allow the use of an informative prior or other prior information until now.[19]
- RIEPS is the regression imputation technique as proposed in principle by Rubin (1986) and discussed earlier in Section 4.4 with an added random residual to create multiple imputations. These multiple imputations are not proper because the step of sampling the parameters from their observed-data posterior distribution is omitted; see Schafer (1999a). Nevertheless, for the matching task we are able to introduce prior information and, moreover, regression imputation is fairly widespread among practitioners; thus RIEPS may serve as the baseline here.

Within NIBAS the prior conditional correlation $\rho_{XY|Z}^{prior}$ is used directly in Step 3 of Algorithm (1). For NORM the unconditional covariance σ_{XY}^{prior} is calculated according to $\sigma_{XY}^{prior} = \sigma_{XY|Z}^{prior} + \widehat{\Sigma}_{XZ}\widehat{\Sigma}_{ZZ}^{-1}\widehat{\Sigma}_{ZY}$ with $\sigma_{XY|Z}^{prior} = \rho_{XY|Z}^{prior}\sqrt{\widehat{\sigma}_{XX|Z}\widehat{\sigma}_{YY|Z}}$ and the covariances being estimated from all available data, i.e., $\widehat{\sigma}_{XX|Z}$ and $\widehat{\Sigma}_{XZ}$ from file A, $\widehat{\sigma}_{YY|Z}$ and $\widehat{\Sigma}_{ZY}$ from file B, and $\widehat{\Sigma}_{ZZ}$ using all values of Z from the n available data sets. Finally, σ_{XY}^{prior} is taken as the starting value for the data augmentation algorithm itself to fix this parameter. Within RIEPS the prior setting of $\sigma_{XY|Z}^{prior}$ is used to retain the correct regression coefficient for the regression of X on Z and Y in file A and Y on Z and Y in file B according to (4.37).[20]

The whole procedure of generating and discarding data, predicting the missing values as described above $m = 5$ times for each generated data set, and calculating the usual multiple imputation estimates $\widehat{\theta}_{MI} = (1/m)\sum_{t=1}^{m}\widehat{\theta}^{(t)}$ of the parameters $\theta = (\mu_X, \mu_Y, \sigma_{XX}, \sigma_{YY}, \rho_{XY})$ for each imputed data set, is repeated $k = 50$ times.[21] The within-imputation

[18]For implementation the S-PLUS library NORM is used here; see Schafer (1996).

[19]The S-PLUS library MICE is used with the default setting of method "impute.pmm", which performs the final imputation on the basis of the predictive mean matching procedure as described in Section 4.4.3.

[20]For a detailed description of the algorithm's implementation, see the S-PLUS routine EVALprior in the appendix.

[21]Note that NORM and NIBAS are computationally rather speedy algorithms, but MICE is not; thus we decided to restrict the repetitions to $k = 50$.

variance $W = (1/m) \sum_{t=1}^{m} \widehat{var}(\widehat{\theta}^{(t)})$ and the between-imputation variance B are calculated $k = 50$ times. According to the MI principle (see Section 4.3.2), we must assume that based on the complete data the point estimates $\widehat{\theta}$ are approximately normal with mean θ and variance $\widehat{var}(\widehat{\theta})$. Therefore it might be reasonable to transform some estimates to a scale for which the normal approximation works well. Since the sample size is quite large here we calculate the usual ML estimates and their asymptotic variances derived from the normal model. From classical statistical theory we know that the ML estimates of μ and σ^2 based on n independent observations are asymptotically normal; i.e.,

$$(\widehat{\mu}_{ML}, \widehat{\sigma}^2_{ML}) \sim (asy)\ N_2\left((\mu, \sigma^2)', I(\mu, \sigma^2)^{-1}\right) \tag{4.62}$$

$$\text{with } I(\mu, \sigma^2)^{-1} = \begin{pmatrix} \sigma^2/n & 0 \\ 0 & 2(\sigma^2)^2/n \end{pmatrix}. \tag{4.63}$$

Moreover, the ML principle provides an asymptotically unbiased estimate of I^{-1} with

$$I(\widehat{\mu}_{ML}, \widehat{\sigma}^2_{ML})^{-1};$$

thus we use as estimates in our study

$$\widehat{\mu}_{ML} = \frac{1}{n}\sum_{i=1}^{n} U_i, \quad \widehat{var}(\widehat{\mu}_{ML}) = \widehat{\sigma}^2/n, \tag{4.64}$$

$$\widehat{\sigma}^2_{ML} = \frac{1}{n}\sum_{i=1}^{n}(U_i - \widehat{\mu}_{ML})^2, \quad \widehat{var}(\widehat{\sigma}^2_{ML}) = 2(\widehat{\sigma}^2_{ML})^2/n, \tag{4.65}$$

for $U = X$ in file A and for $U = Y$ in file B. The 95% MI interval estimates are calculated as is (4.26), with $\widehat{\theta}_{MI} \pm \sqrt{T}\, t_{0.975,\nu}$, $T = W + (1 + m^{-1})B$, enabling us to count the coverage, i.e., the number of times out of k that cover the true population parameter θ.

The sampling distribution of Pearson's product moment correlation coefficient

$$\widehat{\rho}_{XY} = \widehat{\sigma}_{XY}/\sqrt{\widehat{\sigma}_{XX}\widehat{\sigma}_{YY}}$$

is known to be skewed, especially if the corresponding correlation coefficient of the population is large. Thus, usually the multiple imputation point and interval estimates of a correlation ρ are calculated by means of the Fisher z-transformation $z(\widehat{\rho}) = 0.5 \ln((1 + \widehat{\rho})/(1 - \widehat{\rho}))$, which makes $z(\widehat{\rho})$ approximately normally distributed with mean $z(\rho)$ and constant variance $1/(n - 3)$ (see, e.g., Schafer (1997), p. 216, and Brand (1999), p. 116). By back-transforming the corresponding MI point and interval estimates of z via the inverse Fisher transformation the final estimates and confidence intervals for ρ are achieved.

Finally the expectation $E(\widehat{\theta}_{MI})$ of each MI estimate as well as the expectation $E(B)$ of the between-imputation variance and the expectation

$E(W)$ of the within-imputation variance are estimated by simply averaging over k repetitions. Moreover the variance $var(\widehat{\theta}_{MI})$ is estimated by the sample variance of the k MI estimates.

4.8.3 Results Based on Prior Information

Not to bother the reader with endless lists, the results are displayed in detail in the appendix. Here we focus on the MI estimate $\widehat{\rho}_{MI} = \widehat{\rho}_{XY}$ of the unknown association of X and Y, i.e., the third and second levels of validity are discussed in the following. Notice that it is meaningless to discuss whether the first level of validity is achieved by any matching technique because the variables are continuous here. Since the data are generated from a normal model and imputed under a normal model the joint distribution of the imputed variables Y and X is expected to be normal again. In this case achieving the second and the third levels of validity may be discussed only by means of preserving the correlation structure between X and Y. Whether the imputation techniques are able to preserve the joint distributions under less simplifying modeling assumptions is addressed in the evaluation study in Chapter 5.

The estimated expectation $\widehat{E}(\widehat{\rho}_{XY})$, the standard error $s(\widehat{\rho}_{XY})$, denoted by

$$s(\widehat{\rho}_{XY}) = \sqrt{\widehat{var}(\widehat{\rho}_{XY})} = \sqrt{\frac{1}{k-1}\sum_{j=1}^{k}(\widehat{\rho}_{XY}^{(j)} - \widehat{E}(\widehat{\rho}_{XY}))^2},$$

and the coverage in percentage from our $k = 50$ repetitions of the (back-transformed) unconditional correlation for each procedure are listed in Table 4.6. In this table we display the results for prior values -0.9, -0.8, -0.4, 0, 0.4, 0.8, and 0.9 of the conditional correlation $\rho_{XY|Z}^{prior}$.

When calculating a t-statistic like value according to $t = \sqrt{k}(\widehat{E}(\widehat{\rho}_{XY}) - \rho_{XY}^{prior})/s(\widehat{\rho}_{XY})$ to ease interpretation, we realize that with NORM and NIBAS only $\rho_{XY|Z}^{prior} = -0.9$ of the settings above yields an absolute value of t greater than three, whereas RIEPS only once under conditional independence has a t-value less than three. The association given a priori is reproduced quite well by the noniterative multivariate Bayesian regression and the data augmentation procedures. The reproduction is a little bit better and the between-imputation variance is a bit smaller with NIBAS. Table 4.6 shows very clearly that using a Bayesian-based imputation method preserves the prior association far better than a pure frequentist one; see for comparison Moriarity and Scheuren (2001).

With MICE conditional independence of X and Y given $Z = z$ is reproduced well as presented in Table 4.7. It seems to be an interesting area for further research to make the MICE library flexible to prior information.

	NORM			NIBAS			RIEPS		
ρ_{XY}^{prior}	$\widehat{E}(\widehat{\rho}_{XY})$	$s(\widehat{\rho}_{XY})$	Cvg.	$\widehat{E}(\widehat{\rho}_{XY})$	$s(\widehat{\rho}_{XY})$	Cvg.	$\widehat{E}(\widehat{\rho}_{XY})$	$s(\widehat{\rho}_{XY})$	Cvg.
0.3098	0.3007	0.0160	0.96	0.3034	0.0139	0.98	0.3186	0.0182	0.76
0.3402	0.3339	0.0240	0.90	0.3405	0.0153	0.94	0.3547	0.0179	0.72
0.4618	0.4588	0.0524	0.80	0.4616	0.0143	0.98	0.4720	0.0135	0.84
0.5833	0.5925	0.0780	0.84	0.5825	0.0088	1.00	0.5835	0.0106	0.98
0.7049	0.6954	0.0609	0.86	0.7048	0.0098	0.98	0.7218	0.0096	0.52
0.8265	0.8330	0.0230	0.80	0.8250	0.0100	0.92	0.8637	0.0076	0.00
0.8569	0.8563	0.0137	0.88	0.8551	0.0089	0.90	0.8833	0.0081	0.00

Table 4.6. Simulation study (NORM, NIBAS, RIEPS) using prior information

		MICE		
$\rho_{XY\|Z}^{prior}$	ρ_{XY}^{prior}	$\widehat{E}(\widehat{\rho}_{XY})$	$s(\widehat{\rho}_{XY})$	Cvg.
0.0	0.5833	0.5751	0.0146	0.96

Table 4.7. Simulation study (MICE) assuming conditional independence

With respect to the "overall coverage" (for details see the tables in the appendix) we find that NORM, NIBAS, and even MICE sustain the marginal parameters quite well. Obviously the fourth level of validity is well achieved by the alternative approaches. RIEPS is less efficient in reproducing the means and even less concerning the variances and the correlation. Obviously the variation introduced by adding a random residual is not enough. The best coverage of 94% is achieved with RIEPS when conditional independence is assumed via prior information; otherwise the mean coverage is only 73%. With NORM the coverage is high throughout with about 94%, and the NIBAS MI interval estimates even cover the true parameter more than 95%. MICE is not comparable here, since we cannot introduce the prior information and thus conditional independence is established automatically. Concerning only the marginal parameters the coverage is over 93%, though.

4.8.4 Results Based on an Auxiliary Data File

Now we make use of a third data source. While setting the true $\rho_{XY} = 0.8$ again we draw a small but complete data set for (z, x, y) according to $(Z, X, Y) \sim N_4(0, \Sigma)$ as given with (2.18). For imputations via NORM and MICE the complete data are simply added to the incomplete data and their imputation procedures are performed based on the usual improper priors. With NIBAS and RIEPS the conditional correlation $\rho_{XY|Z}$ is estimated by means of a small sample first. Therefore we estimate σ_{XY} from this third data source and $\widehat{\sigma}_{XY|Z}$ according to $\widehat{\sigma}_{XY|Z} = \widehat{\sigma}_{XY} - \widehat{\Sigma}_{XZ}\widehat{\Sigma}_{ZZ}^{-1}\widehat{\Sigma}_{ZY}$. Then $\widehat{\rho}_{XY|Z}$ is derived by $\widehat{\rho}_{XY|Z} = \widehat{\sigma}_{XY|Z}/\sqrt{\widehat{\sigma}_{XX|Z}\widehat{\sigma}_{YY|Z}}$ and used as the prior in Step 3 of Algorithm (1) with NIBAS or to calculate the regression coefficient for RIEPS according to (4.37).[22]

Table 4.8 displays the mean estimate of the unconditional correlation and further statistics when samples of 1 and 5% of the actual sample are available to improve the imputation procedure. For the estimation process itself we used the z-transformation and transformed the estimates back afterwards; thus the within-imputation variance W is constant and therefore is not listed in Table 4.8. From each of the k matched data sets we calculated the MI estimate of ρ_{XY} and stored its between-imputation variance.

[22]For details of the algorithm and its calculations see the S-Plus routine EVALcf in the appendix.

n	Procedure	Sample	NORM	NIBAS	MICE	RIEPS
50	$\widehat{E}(\widehat{\rho}_{XY})$	0.8014	0.7958	0.7985	0.7984	0.8359
	$s(\widehat{\rho}_{XY})$	0.0583	0.0272	0.0255	0.0215	0.0278
	$\min(\widehat{\rho}_{XY})$	0.6157	0.7092	0.7252	0.7426	0.7463
	$\max(\widehat{\rho}_{XY})$	0.8995	0.8389	0.8403	0.8372	0.8726
	$\widehat{E}(B)$	–	0.0016	0.0002	0.0027	0.0001
	Cvg.	0.90	0.90	0.56	0.88	0.10
250	$\widehat{E}(\widehat{\rho}_{XY})$	0.8001	0.8007	0.8001	0.8000	0.8386
	$s(\widehat{\rho}_{XY})$	0.0231	0.0096	0.0109	0.0092	0.0106
	$\min(\widehat{\rho}_{XY})$	0.7166	0.7784	0.7669	0.7777	0.8031
	$\max(\widehat{\rho}_{XY})$	0.8571	0.8178	0.8233	0.8177	0.8595
	$\widehat{E}(B)$	–	0.0006	0.0003	0.0006	0.0001
	Cvg.	0.96	0.94	0.88	0.92	0.02

Table 4.8. Simulation study using a third data source

Based on the small samples of size $n = 50$ and $n = 250$ the z-transformed Pearson's product moment correlation coefficient and its asymptotic confidence interval $z(\widehat{\rho}) \pm \lambda_{0.975}\sqrt{1/(n-3)}$ using the 97.5% quantile $\lambda_{0.975}$ of the standard normal were calculated as well. After back-transforming the point and interval estimates from the small samples the final estimates were derived. Finally, from these k estimates again the estimated expectation, its standard error, the minimum and maximum values, and the coverage (in percent) based on the 95% confidence intervals were computed.

Clearly we see that even a very small sample of size $n = 50$ is suitable for substituting the arbitrary prior used before and, thus, improves the imputation procedure. It is worth taking all the information provided by the two files into account rather than basing the estimation on the small third file alone. The range of the correlation estimate derived from the small 1% sample will be considerably narrowed from $\widehat{\rho}_{XY}^{sample} \in [0.6157 \quad 0.8995]$ to the smaller intervals of $\widehat{\rho}_{XY}^{NORM} \in [0.7092 \quad 0.8389]$, $\widehat{\rho}_{XY}^{NIBAS} \in [0.7252 \quad 0.8403]$ and $\widehat{\rho}_{XY}^{MICE} \in [0.7426 \quad 0.8372]$ by using the proposed procedures. For MICE the number of iterations is set to 250 (150) for the $n = 50$ (250) auxiliary file leading to a run-time of about three hours for each simulation run on an AMD Duron 750 MHz computer. The long run-time with MICE is necessary as otherwise there is the tendency to impute X and Y values towards conditional independence. RIEPS precisely overestimates the true correlation with a negligible between-imputation variance. Now the coverage is best with NORM. NIBAS produces rather small MI interval estimates due to its rather small between-imputation variance.

4.8.5 Summary

According to Rubin (1996) a proper imputation method requires that the multiple imputation estimates $(\widehat{\theta}_{MI}, W)$ are unbiased estimates of the statistics $(\widehat{\theta}, \widehat{var}(\widehat{\theta}))$ of the hypothetical complete data set. Furthermore the extra inferential uncertainty about θ due to missing data should be correctly reflected; i.e., $var(\widehat{\theta}_{MI}) = (1 + m^{-1})B$ should hold theoretically.

Brand (1999), pp. 104–116, demonstrates in detail the validation of methods for the generation of imputations. Roughly speaking, a proper MI method may be verified if the averaged MI estimates $\widehat{E}(\widehat{\theta}_{MI})$ and their averaged within-imputation variance $\widehat{E}(W)$ are approximately equal to the estimate $\widehat{\theta}$ and its variance estimate $\widehat{var}(\widehat{\theta})$ when based on the actual complete data. Furthermore, the averaged between-imputation variance $(1 + m^{-1})\widehat{E}(B)$ should be approximately equal to the sample variance $\widehat{var}(\widehat{\theta}_{MI})$ of the MI estimate from the repeated simulations. As far as available, these values are listed in the tables of the appendix. The judgment, however, as to whether two quantities are approximately equal is still left to an intuitive criterion (see Brand (1999), p. 115) and is beyond our interest.

Since all estimates used herein are at least asymptotically unbiased and the sample size is quite large, the coverage of the MI interval estimate gives a good hint as to whether the implemented procedure may be proper. If the multiple imputations are proper, the actual interval coverage should be equal to the nominal interval coverage. Concerning the marginal distributions NORM, NIBAS, and MICE are apparently proper but prior information of ρ_{XY} is only reproduced suitably by NORM and NIBAS. RIEPS provides the lowest coverage due to the fact that variances are often under- and correlations overestimated.[23] Whether the parameters of the model are sampled from their complete or observed-data posterior distribution, the extra amount of uncertainty induced thereby improves the validity of the imputation techniques considerably. According to the prior information that may be used, NORM and NIBAS seem to achieve the fourth as well as the third (and second) level of validity.

A third data source is best exploited by NORM and MICE here, since the between-imputation variance based on NIBAS is small throughout. Thus the MI confidence intervals are smaller there leading to less conservative tests.

[23]This general effect of regression imputation has been shown in other simulation studies as well; see, e.g., Nordholt (1998) and Rässler (2000).

4.9 Concluding Remarks

By means of the simulation study we have realized that RIEPS is not a proper imputation method even if the data follow simplifying assumptions, if, for example, the data are generated according to the data model assumed. Besides some theoretical shortcomings MICE has its disadvantages concerning speed and utilizing prior information. If the normal model fits to the data or prior assumptions about the association of the variables never jointly observed other than conditional independence are suitable, we propose to use NORM or NIBAS for the imputation process; otherwise MICE is a rather flexible alternative at hand.

In the matching task which can be viewed as a very special missing data pattern, the observed-data posterior distribution is derived under the assumption of a normal data model. Thus, a particular model specially suited for the matching task is postulated. The assumption of normality seems to be a great limitation at first glance, but at present we have not yet found other parametric models, for example, mixture models or conditional Gauss models,[24] that can be fitted to data sets with hundreds of variables, displaying the multivariate structure of the data as well as meeting the requirements of the statistical matching task. Moreover, the imputer's model is designed to create meaningful imputations preserving the structure of the data but not used for inference itself. In Chapter 5 we discuss some solutions concerning mixtures of variables.

The iterative univariate imputation procedure introduced in Section 4.7 tries to reduce the problem of dimensionality to multivariate regressions with univariate responses. It is a very flexible procedure allowing different scales for the variable of interest but at the expense of controlling the results. Prior information cannot be used efficiently here.

Another great advantage of the alternative approaches proposed herein we find in the property of multiple imputations to reflect the uncertainty due to the missing data. MI confidence intervals are typically wider than the conventional ones; thus tests based on multiply imputed data sets are conservative. If a null hypothesis is rejected due to MI tests, there is strong evidence that the underlying data do not correspond to the null hypothesis although the amount of missing information may be high. Moreover, we have seen that within MI it is possible to estimate the bounds of the unconditional association of the variables never jointly observed simply by using different prior settings.

Comparing the number of parameters that have to be drawn from their particular posterior distributions it is shown that NIBAS requests the

---

[24] We refer to the applications presented by Schafer (1997), Gelman et al. (1998), Raghunathani and Grizzle (1995), Kamakura and Wedel (1997), and Cox and Wermuth (1996). They are either based on normal models or limited in its variable space.

fewest parameters. We should carefully note that the number k of common variables Z for NIBAS and MICE includes the constant, however, this in not true for NORM. Considering the simulation study with univariate X and Y and two common variables Z we get $4 \cdot 5/2 + 4 = 14$ parameters to be calculated by NORM, $2 \cdot 5 = 10$ parameters by MICE, and $2 \cdot 3/2 + 2 \cdot 3 = 9$ parameters by NIBAS. A recently performed match of TV viewing behavior data with radio data in Germany called MA2000 (see Wendt (2000)) has 128 common variables Z, some being continuous, about 120 continuous variables X, and about 100 continuous variables Y. Thus we find ourselves confronted with performing random draws for a 348×348 matrix with 60726 distinct elements and 348 means, i.e., 61074 parameters when NORM is used. With MICE there are $349 \cdot 220 = 76780$ parameters to draw randomly. Finally, NIBAS requests random draws for squared matrices of sizes 100 and 120, respectively. A total of 52470 parameters has to be operated by NIBAS. It is easy to show in general that NIBAS has the smallest number of parameters, although it requires random draws from higher-dimensional normal distributions than the other procedures, when $pk > r$ or $qk > r$ holds.[25] Since computational power is increasing rapidly we regard these limitations to be of minor importance.

Some alternative approaches have been discussed recently in the context of missing values when the number of variables is large. Song and Belin (1999) propose a factor analysis approach to reduce the problem of dimensionality, viewing the data as being consistent with a smaller number of uncorrelated factors comprised of variables that are correlated within factors. They derive a multiple imputation procedure based on the complete-data posterior distribution of the model parameters and a Gibbs sampling routine. Thus their model is iterative again. Unfortunately the parameterization is quite difficult in terms of hyperparameters Λ impeding the application of their approach to the matching task for the moment. However, this may be an interesting area for further research.

Furthermore, we have seen that prior information is most easily used by NIBAS whereas with RIEPS the variation of the variables being imputed is too small. MICE does not allow the use of prior information yet. With NORM prior information can only be applied via the hyperparameter when the standalone MS-Windows^{TM} version is used. In the simulation study the auxiliary file is most efficiently used by NORM and MICE.

To proceed further, we may loosen the assumption of no data missing within the single data sets, an assumption we made at the beginning. Usually each single data source will have some missing data due to nonresponse.

[25] With the usual S-PLUS software for MS-Windows^{TM} random draws from a, e.g., 1000-dimensional multivariate normal distribution can be performed easily on a conventional personal computer. Depending on the computer used for the calculations and its memory, sometimes the S-PLUS object size has to be set up to enable the computation of these high-dimensional normally distributed random variables.

Assuming a MAR mechanism we propose to first impute these missing values for each single data file, for example, with a suitable data augmentation procedure. After these gaps are filled the quick algorithm NIBAS may be used to realize the "mass imputation" due to the statistical matching task.

5

Empirical Evaluation of Alternative Approaches

In a situation where social scientists are so hungrily looking for increasingly rich data bases, statistical matching is a dangerously attractive procedure for creating files containing the logical union of the variables involved in either of the component files [...] I would like to see a good deal of empirical evaluation of the validity of such joint distributions before I would suggest removing the label from this procedure: "DANGEROUS – USE WITH CAUTION".

I.P. Fellegi (1977).

5.1 Introduction and Chapters Outline

In this chapter we consider a typical European statistical matching task where multiple categorical, continuous, and so-called semicontinuous variables concerning media and television behavior have been recorded in separate surveys in addition to the usual demographic and socioeconomic information. Our goal here is less to analyze or describe the relationship among variables in a meaningful way but to find out whether the procedures for imputing missing values preserve important features of marginal and joint distributions. We want to investigate the performance of the proposed alternative matching techniques discussed in Chapter 4 when applied to real data sets which typically do not follow simplifying assumptions. The validity of a matching technique is measured according to the four levels we introduced in Chapter 2. The data are provided from the television behavior panel run by the largest German market research company.[1] This data set is matched regularly with the purchasing behavior panel; the procedure actually applied by the GfK is described in section 3.3.5.

During 1998 some data were taken from this panel survey which had been fully observed. Thus we got access to nearly 5500 observations with a variety of variables whereof we take 30 variables to use for the evaluation study. Eight variables describe demographics of the people and the households sampled, another six variables contain the averaged TV viewing time of some genres, seven variables give the (estimated) probabilities of viewing a specific channel, and nine variables display the consuming behavior. In Chapter 3 we explained this situation as a very typical setting for media targeting in Europe. For illustration purposes the number of variables is restricted to 30.

This chapter begins with a description of the data and practical transformations to produce the necessary model assumptions are discussed. The media data used in this chapter are heavily skewed and have large proportions of zero values. Some metrics and statistics are introduced to find a suitable measurement to assess the degree of validity a matching technique may achieve. With this real data set a comprehensive simulation study is performed and the alternative Bayesian approaches NIBAS, NORM, and MICE already discussed in Chapter 4 are applied. Notice that we do not make any inference about the contents of the data; our focus is on discussing the performance of these matching techniques when real data are to be matched. Hence we compare inferences drawn from the complete original data with inferences based on the imputed data, i.e., the statistically matched files. For the imputed data set inference is derived according to the multiple imputation principle. Finally, additional information derived from auxiliary files of different sizes is also used. We want to show that the alternative approaches are able to efficiently and easily use auxiliary information.

In the survey data set the amount of specific variables being conditionally independent given the common variables is already rather high. Therefore, we recalculate the investigations with generated data sets having parameters similar to the survey data set but with higher conditional associations. So every analysis in the evaluation study is twofold: the real survey data and the simulated data used for it. Matching techniques such as the Bayesian approaches are model-based; thus we also investigate the influence of departures from the assumption of normality as well as from the assumption of conditional independence.

[1]The German Institute for Market Research, GfK AG, kindly supported this research.

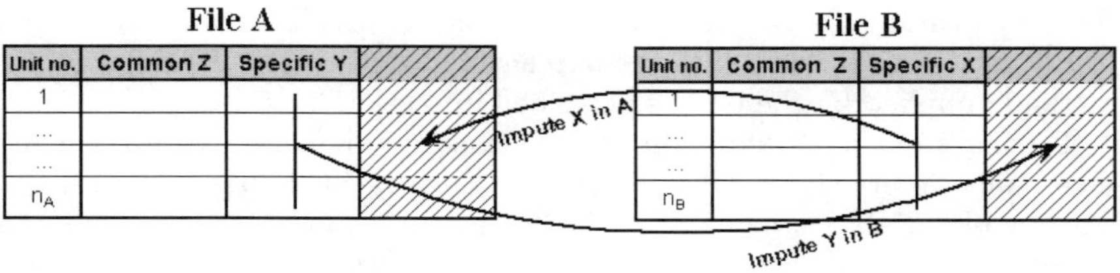

Figure 5.1. Imputed variables in the empirical evaluation

5.2 Simulation Study Using Survey Data

The survey data have a set of 14 variables Z, which we regard as the common variables. These 14 variables contain information about gender, age, and size of household as well some information about the time particular TV programs have been seen. Adding such information to the common variables as so-called common specific variables is an approach published by Roberts (1994) and quite successfully carried out by the German market research company GfK AG; see Section 3.3.5.

Then seven specific variables X describe the TV viewing behavior and nine specific variables Y describe the purchasing behavior of the $n = 5496$ individuals. Initially there are no missing values at all in this complete data matrix of size 5496×30. Two data sets A and B are created from this survey data. File A contains units $1, 2, \ldots, n_A$ with data corresponding to common variables Z and specific variables Y; variables X are set as missing. The other file B consists of units $1, 2, \ldots, n_B$ which is equivalent to $n_A + 1, n_A + 2, \ldots, n$ with their data corresponding to common variables Z and specific variables X with variables Y missing. Thus, the variables X and Y are never jointly observed in this setting. With these two data sets we want to investigate the power of matching techniques when first X is imputed in file A and second, Y is imputed in file B as Figure 5.1 illustrates.

In Chapter 2 we show that under some frequentist criteria the univariate distributions of the imputed variables are preserved at least by most of the traditional approaches. Thus, we define achieving the fourth level of validity as a minimum requirement of any matching technique. According to this requirement the reproduction of the marginal distributions is discussed first. Furthermore, the third level of validity according to our terminology is investigated, i.e., sustaining the correlation structure by analyzing the conditional associations of X and Y given $Z = z$ as well as the unconditional associations of X and Y. To analyze the second level of validity, the multivariate distributions of X and Y in the complete data are compared with the multivariate distributions of X and Y in the imputed data after the matching from B to A and A to B, respectively. Due to the variables being continuous there is no need to discuss the first level of validity; counting hits is meaningless here.

Many linear regression analyses are performed to investigate and compare the significance of coefficients in the complete data set and the imputed data set. In our opinion a satisfying matching technique should yield (more or less) the same conclusion for any typical significance level within the matched data. This investigation also aims at achieving the third level of validity because the conditional correlations are incorporated in the regression coefficients.

Moreover, samples of size $n = 300$ and $n = 600$ are drawn from the original data which are regarded as a completely observed data source. We call this third source of information an auxiliary file throughout this work. Such auxiliary files may be available to the research companies, if they are willing to invest the time and the money to collect such data. A third source of information about the variables X and Y never jointly observed in the large files could also represent a sample drawn at an earlier point in time than the other files, or come from a different population, or from a sample of inferior quality. Consequently Rodgers (1984) states that the interpretation of analyses based on the matched files having used such a (biased) auxiliary file in any way, should be especially cautious. Thus, we explore the extent to which a small but completely observed file (i.e., the auxiliary file) may improve or mislead the statistical inference drawn from the matched data set.

5.2.1 Univariate Structure of the Data

Original Data Set and Empirical Moments

The original and complete data set of size $n = 5496$ shows the following structure displayed in Table 5.1. The first three common variables Z_1, Z_2, and Z_3 are binary coded; Z_4 and Z_5 are of categorical type. All other variables are more or less continuous. From Z_9 to Z_{14} the common variables contain the average of seconds a TV series has been seen by an individual during some time period. The X variables measure the averaged relative frequencies a unit is using a specific TV channel during a particular time period. The Y variables count the average number of buys of a consumer good during a particular time period. The range of each variable, its mean, and standard error (se) are displayed in Table 5.1.

Notice that usually survey data contain weights that add up to the size of the population. In German market research, e.g., the individual weights often are calculated according to the distribution of some sociodemographic variables in the population, such as gender, age groups, or other criteria; see Koschnick (1995), pp. 703–705. However, in the following investigations we treat the actual data and the statistically matched (imputed) data as two simple random samples from any population of interest. Inferences

i	Variable	Min	Max	% Zeros	Mean	Se
Z_1	Gender	0	1	0.18	0.82	0.38
Z_2	West/East	0	1	0.25	0.75	0.44
Z_3	Fully employed	0	1	0.70	0.30	0.46
Z_4	Hearing pop music	1	5	-	3.17	1.69
Z_5	Hearing folk music	1	5	-	3.36	1.60
Z_6	Size of household	1	7	-	2.40	1.23
Z_7	Number of children	1	5	-	1.39	0.77
Z_8	Age of person	17	94	-	49.65	15.23
Z_9	Medical series	0	3563.71	0.11	94.09	237.95
Z_{10}	Daily soaps	0	5897.92	0.15	126.85	364.89
Z_{11}	News	0	5905.28	0.01	825.10	719.04
Z_{12}	Business	0	660.81	0.03	50.28	68.08
Z_{13}	Show	0	10291.65	0.01	433.52	581.72
Z_{14}	Classic	0	1453.33	0.07	19.72	52.84
X_{15}	Super RTL	0	1	0.31	0.09	0.14
X_{16}	Kabel1	0	1	0.20	0.17	0.21
X_{17}	Pro Sieben	0	1	0.12	0.28	0.26
X_{18}	RTL2	0	1	0.14	0.18	0.19
X_{19}	RTL	0	1	0.05	0.42	0.28
X_{20}	ZDF	0	1	0.02	0.43	0.25
X_{21}	ARD	0	1	0.01	0.49	0.25
Y_{22}	Margarine	0	97.62	0.07	10.72	10.83
Y_{23}	Cola drinks	0	983.00	0.22	33.66	77.78
Y_{24}	Desserts	0	146.57	0.27	6.22	9.88
Y_{25}	Butter	0	80.75	0.05	8.22	8.56
Y_{26}	Ketchup	0	36.30	0.37	1.38	2.48
Y_{27}	Coffee	0	97.75	0.09	8.95	8.14
Y_{28}	Frozen pizza	0	30.64	0.38	1.80	3.40
Y_{29}	Dessert powder	0	6.01	0.32	0.04	0.27
Y_{30}	Cake mixtes	0	65.50	0.32	1.55	3.83

Table 5.1. Structure of the original media data set

based on these two data sets – the true and the statistically matched one – are compared. The underlying population is never the aim of our analysis. Therefore, we discard the original weights for the following simulation study.

Univariate Distributions of the Variables to be Imputed

The proportions of zeros are nearly up to 40% for some purchasing and viewing variables. This requires great care applying the model-based matching techniques to the data. Moreover, the data do not satisfy any typical distributional assumption. Since the model-based matching techniques introduced in Chapter 4 usually assume multivariate normality at least for the variables to be imputed, we discuss several transformations of the data later in Section 5.2.1.

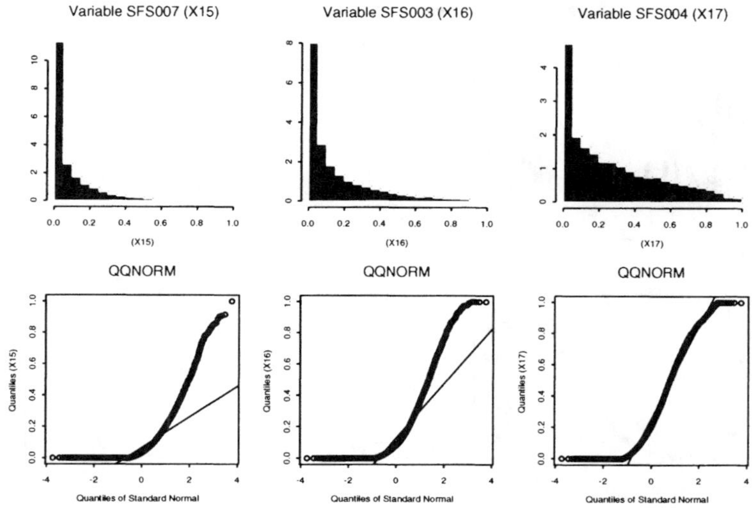

Figure 5.2. Histogram plots and qqplots for variables X_{15}, X_{16}, and X_{17}

Figure 5.2 shows the exemplary histogram plots for the first three variables of X, i.e., X_{15}, X_{16}, and X_{17}. The quantile–quantile plots, so-called "qqplots" or "qqnorm" plots, presented here compare the quantiles of the estimated cumulative density function (c.d.f.) of each X variable with the quantiles of the standard normal cumulative density function.

The qqplots are used to assess whether data have a particular distribution, or whether two data sets have the same distribution. If the distributions are the same, then the plot will be approximately a straight line, which is also indicated in Figure 5.2.[2] Most of our survey variables are severely skewed due to their proportions of zeros.

The X variables are bounded to [0,1], whereas the Y variables have support on the positive axis with different maximum values as Figure 5.3 points out for the first three variables of Y, i.e., Y_{22}, Y_{23}, and Y_{24}.

Clearly, any test of the hypothesis of univariate normality will be rejected with these data. For the Bayesian approaches (NORM and NIBAS) we assume a normal distribution at least for the variables to be imputed, only the iterative univariate imputation procedure (MICE) allows us to use more flexible distributions of the imputed variables. Thus, we have to perform some transformations first, before applying these matching techniques. Theoretically, we should transform the data to achieve mul-

[2]The extreme points have more variability than points toward the center. A plot from S-PLUS function qqnorm that is bent down on the left and bent up on the right means that the observed data have longer tails than normal data have.

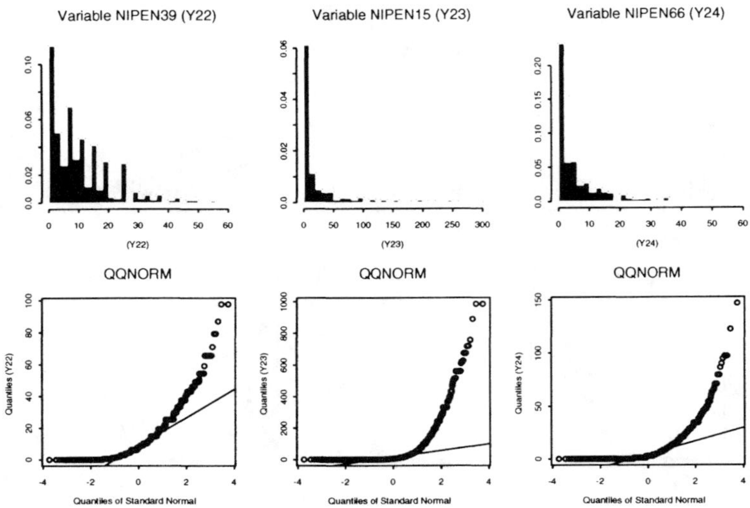

Figure 5.3. Histogram plots and qqplots for variables Y_{22}, Y_{23}, and Y_{24}

tivariate normality. Practically, such transformations are not yet available; the usual transformations are performed on a univariate scale (see, e.g., Schafer (1997), p. 212).

Transforming the Data for Imputation Purposes

Some of the matching techniques mentioned earlier are model-based and require at least more or less univariate normally distributed data. An ordinary log transformation or anything similar is not directly applicable because the data have proportions of zeros too large to simply be ignored. Thus, we cut those variables into two. Note that the following transformations are done only for the sake of imputation, since there are neither proper multi- nor univariate transformations on the undivided data these transformations do not improve inference, which is discussed in the forthcoming sections.

To proceed further, we treat the majority of variables as being semicontinuous, with a proportion of responses at the fixed value of zero and a continuous distribution among the remaining variables. Subject to an approach published by Schafer and Olsen (1999), one may encode each semicontinuous variable U to a binary indicator W (with $W = 1$ if $U \neq 0$ and $W = 0$ if $U = 0$) and a continuous variable V which is treated as missing whenever $U = 0$; for an illustration, see Figure 5.4.

Notice that a relationship between W and V would have little meaning and could not be estimated by the observed data. However, we aim at generating plausible imputations for the original semicontinuous variable U and, thus, are only interested in the marginal distribution for W and

Unit no.	U
1	12
2	NA
3	0
4	0
5	NA
...	...
n_A-1	3
n_A	0

⇨

Unit no.	W	V
1	1	12
2	NA	NA
3	0	NA
4	0	NA
5	NA	NA
...		...
n_A-1	1	3
n_A	0	NA

Figure 5.4. Encoding semicontinuous variables

the conditional distribution for V given $W = 1$. EM and MCMC procedures have been shown to behave well in this context with respect to the parameters of interest.[3]

We encode the variables X_i to $W_i = 1$ and $V_i = g(X_i)$ if $X_i \neq 0$ and $W_i = 0$ and V_i missing if $X_i = 0$ for $i = 15, 16, \ldots, 21$. Since the values of the variables X are bounded between zero and one representing probabilities, a conventional logit-transformation (see Greene (1997)) seems to work quite well for $g(X)$:

$$g(X) = \left(\frac{X}{1-X}\right) \quad \text{for } X \in (0, 1)\,. \tag{5.1}$$

Every variable X contains one unit with an observed value of $X = 1$; these are treated as $X = 0.999$.

The effect of this split and the logit transformation is shown in Figure 5.5 plotted only for the part of V where $W = 1$.

Likewise we encode the variables Y_j to $W_j = 1$ and $V_j = g(Y_j)$ if $Y_j \neq 0$ and $W_j = 0$ and V_j missing if $Y_j = 0$ for $j = 22, 23, \ldots, 30$. For the positive Y values an ordinary log transformation is a good choice, with

$$g(Y) = \ln(Y)\,. \tag{5.2}$$

We performed other transformations too, especially the Box–Cox transformation with $g(Y) = (Y^\theta - 1)/(\theta)$ for $\theta \neq 0$, but the best normal approximation we found with a value of θ near zero, which indicates a simple log transformation to be the best choice.

Note that if the data are imputed on a log scale and transformed back to the original scale, the imputed values will never be negative. The effect of the split and the log transformation is shown in Figure 5.6, plotted only for V where $W = 1$.

The transformed variables now take negative as well as positive values but the distributions still are skewed. Performing tests for univariate normal

---

[3]For a further description of this unidentified normal model see Schafer and Olsen (1999).

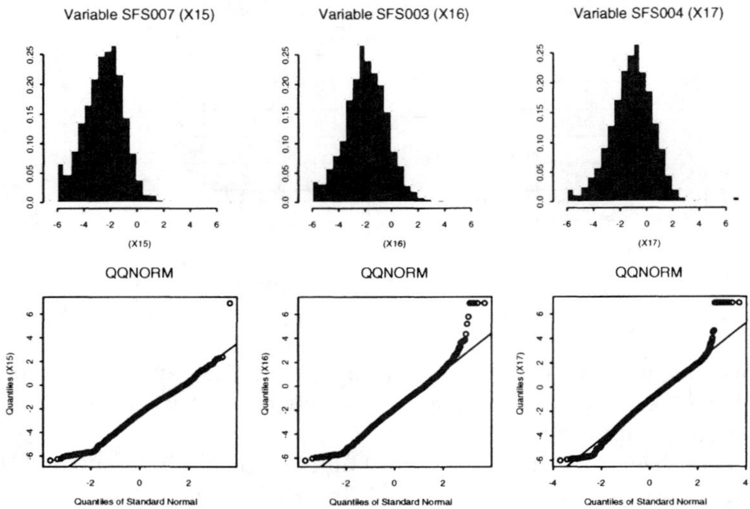

Figure 5.5. Histogram and qqplot for transformed variables X_{15}, X_{16}, and X_{17}

distributions on the transformed parts such as the Kolmogorov–Smirnov test, the χ^2 goodness-of-fit with grouped data, and the Jarque–Bera test will still reject the hypothesis of normality at any reasonable significance level.

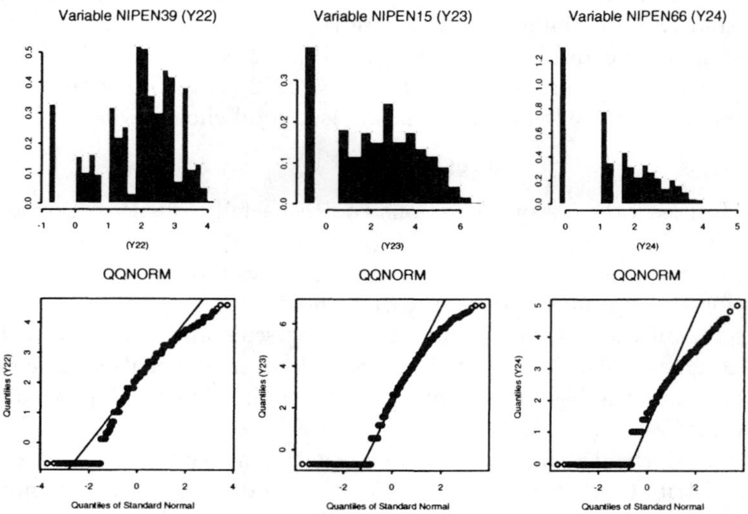

Figure 5.6. Histogram and qqplot for transformed variables Y_{22}, Y_{23}, and Y_{24}

However, common sense shows that such deviations from normality should not harm the imputation process too much; see Schafer (1997), p. 212 or Gelman et al. (1998). A growing body of evidence supports the claim to use a normal model to create multiple imputations even when the observed data are somewhat non-normal. The focus of the transformations is rather to achieve a range for continuous variables to be imputed that theoretically have support on the whole real line than to achieve normality itself. Even for populations that are skewed or heavy-tailed, the actual coverage of multiple imputation interval estimates is reported to be very close to the nominal coverage; see Schafer (1997), p. 212. The multiple imputation framework has been shown to be quite robust against moderate departures from the data model. Caution is required if the amount of missing information is very high, i.e., beyond 50%. Having cut the complete data set into halves, we do not exceed this amount of missingness. Thus, we may proceed further with these transformed data.

For the imputation process itself, the Bernoulli variables Z_1, Z_2, and Z_3 are also treated as normal. This is rather an ad hoc procedure than theoretically sound, but Schafer (1997), pp. 212-218 states that it has worked well in several simulation studies especially when these variables have no missing data. The continuous imputed values for W are rounded to zero or one. Schafer and Olsen (1999) report that the cutoff criterion of 0.5 works well when the overall proportion of zeros is not too large as is the case in this setting. Having finished the imputation process with the transformed data, as a final step, imputations for X and Y are obtained by setting $X_i = 0$ if $W_i = 0$ and $X_i = g^{-1}(V_i)$ if $W_i = 1$ for $i = 15, 16, \ldots, 21$, and analogously by setting $Y_j = 0$ if $W_j = 0$ and $Y_j = g^{-1}(V_j)$ if $W_j = 1$ for $j = 22, 23, \ldots, 30$.

Finally, it is important to recall that a normal distribution has support on the whole real line, but the continuous variables in this data set have a limited range of possible values. Even after having the Y variables transformed to a log scale the imputation might occasionally result in imputed values out of range. Thus, when transforming them back some values will be unrealistically large. To handle this problem we include consistency checks in our imputation routines. After performing the final imputation step for the missing Y values, each row of the imputed data set is examined to see whether any of the imputed values is out of range. In that case the missing data are redrawn until our constraints of $\max(Y) = (200, 2000, 300, 200, 100, 200, 70, 15, 160)'$ are satisfied. According to Schafer (1997), p. 204, this procedure leads to approximate proper (multiple) imputations under a truncated multivariate normal model.

5.2.2 Bivariate Associations of the Original Data

Above all we want to show that the alternative approaches are able to efficiently use prior information even for practical matching tasks. For ex-

r_P	Y_{22}	Y_{23}	Y_{24}	Y_{25}	Y_{26}	Y_{27}	Y_{28}	Y_{29}	Y_{30}
X_{15}	0.07	0.07	0.04	-0.03	0.10	**0.02**	0.03	**0.01**	0.06
X_{16}	0.06	0.05	**0.01**	-0.08	0.06	**0.01**	**0.00**	**0.01**	**0.01**
X_{17}	0.10	0.14	0.09	-0.08	0.16	0.03	0.11	**0.02**	0.04
X_{18}	0.08	0.11	0.06	-0.08	0.11	**0.02**	0.08	**0.02**	0.04
X_{19}	0.10	0.09	0.06	-0.07	0.10	0.05	0.06	**0.01**	0.07
X_{20}	-0.09	-0.14	-0.12	-0.09	-0.14	-0.08	-0.12	-0.04	-0.09
X_{21}	-0.09	-0.13	-0.10	-0.08	-0.13	-0.09	-0.11	-0.05	-0.10

Table 5.2. Correlation structure of the media data

ample, the conditional independence assumption can be made a priori and is not achieved as a matter of fact as the traditional approaches do. To evaluate the association of the (untransformed or back-transformed) variables X and Y in the original sample, suitable tests are performed. However, testing whether the unconditional bivariate association for any two variables Y and X is zero leads to the decision either to ignore the non-normality of the data or to use nonparametric alternatives sacrificing information by reducing the scale. A detailed discussion to find suitable test procedures is given in the following sections.

Pearson's Correlation Coefficient

The correlation structure of the specific variables X and Y estimated by the conventional sample correlation coefficient, i.e., Pearson's product moment correlation coefficient,

$$r_P = \widehat{\rho}(X_i, Y_j) = \frac{\sum_{k=1}^{n}(X_{ik} - \overline{X}_i)(Y_{jk} - \overline{Y}_j)}{\sqrt{\sum_{k=1}^{n}(X_{ik} - \overline{X}_i)^2 \sum_{k=1}^{n}(Y_{jk} - \overline{Y}_j)^2}}, \tag{5.3}$$

$i = 15, 16, \ldots, 21$, and $j = 22, 23, \ldots, 30$ is shown in Table 5.2.

Assuming normality of the data first, we may test for the unconditional bivariate associations applying the S-PLUS function cor.test, which uses the standard test procedure

$$t = \frac{(\sqrt{n-2})\, r_P}{\sqrt{1 - r_P^2}} \quad \sim \quad t(n-2)\,, \tag{5.4}$$

where t under the null hypothesis of $\rho = 0$ is t distributed with $n - 2$ degrees of freedom; see Stuart and Ord (1991), p. 985. To account for the non-normality of the data one may calculate the z-transformation

$$z(r_P) = \frac{1}{2} \log\left(\frac{1 + r_p}{1 - r_P}\right) \quad \sim \quad (\text{appr.})\ N\left(z(\rho), \frac{1}{n-3}\right) \tag{5.5}$$

which makes $z(r_P)$ approximately normally distributed; see Stuart and Ord (1991), p. 984. Hence we calculate the standardized z-values according to $(z(r_P) - 0)/\sqrt{1/(n-3)}$ for $n = 5496$ which leads to Table 5.3.

	Y_{22}	Y_{23}	Y_{24}	Y_{25}	Y_{26}	Y_{27}	Y_{28}	Y_{29}	Y_{30}
X_{15}	5.24	5.46	3.09	-2.52	7.45	1.65	2.09	1.03	4.68
X_{16}	4.57	3.50	0.96	-6.02	4.39	0.99	0.33	0.96	0.72
X_{17}	7.20	10.37	6.97	-5.80	11.66	1.97	7.85	1.55	3.33
X_{18}	5.61	8.38	4.45	-5.64	8.28	1.62	5.91	1.48	3.11
X_{19}	7.57	6.76	4.34	-5.26	7.49	3.50	4.27	0.42	5.43
X_{20}	-6.35	-10.12	-8.78	-6.46	-10.33	-6.10	-8.98	-3.33	-7.00
X_{21}	-6.87	-9.41	-7.76	-5.72	-9.64	-6.42	-8.22	-3.97	-7.09

Table 5.3. Transformed correlations $z(r_P)\sqrt{5493}$ of the media data

If we compare those values above against the ±97.5% quantile (i.e., $\pm\lambda(0.975) = \pm 1.96$) of the standard normal distribution, we find 11 associations being zero at a significance level of 5%, marked with bold letters in tTable 5.2. A further 29 correlations are regarded as positive and 23 as negative.

However, before going ahead we check the efficiency of the z-approximation because our data are very skewed and have large proportions of zeros. Thus we perform a small simulation taking $k = 100$ simple random samples s_k of size $n = 2000$ each [4] without replacement from our complete population of size $n = 5496$. From every sample s_k we calculate the 63 sample correlation coefficients $r_{ij}^{(k)}$, $i = 15, 16, \ldots, 21$, $j = 22, 23, \ldots, 30$ and store their values of the z-transformation $z(r_{ij}^{(k)})$ for $k = 1, 2, \ldots, 100$. Some univariate normality tests are performed on these simulated values. Although we do not have a best test of normality, a rather differentiating test concerning the distributional shape seems to be the χ^2 goodness-of-fit test comparing the observed cell counts with their expected values under the null hypothesis. Clearly, grouping the data sacrifices information. We use 13 cells here to ensure that the expected cell counts are at least 5, and compare the test statistic with the χ^2 distribution with $13-1-2 = 10$ degrees of freedom.[5] In our case the best discriminatory power shows the χ^2 test. Other goodness-of-fit tests such as the Kolmogorov–Smirnov test and the Jarque–Bera test tend to never or always reject the null hypothesis. Table 5.4 displays the p-values for the test of normality of the empirical density function (e.d.f.) $\widehat{f}_{z(r_P)}$.

In only 5 of the 63 cases would we reject the hypothesis of normality for the empirical density function $\widehat{f}_{z(r_P)}$ at a significance level of 5%, explicitly only in one case. The worst two approximations of the normal distribution by the empirical $\widehat{f}_{z(r_P)}$ are presented as qqnorm plots in Figure 5.7.

[4]When discussing the effect of the matching techniques for file A and file B, respectively, the maximum sample size is even larger with $n_A = n_B = 2748$.

[5]We apply the usual procedure and estimate the parameters from the original, i.e., not grouped, data and then subtract one degree of freedom for every parameter estimated.

P-values	Y_{22}	Y_{23}	Y_{24}	Y_{25}	Y_{26}	Y_{27}	Y_{28}	Y_{29}	Y_{30}
X_{15}	0.7386	0.3978	0.1256	0.1350	0.3765	0.1085	0.2642	0.1085	0.1910
X_{16}	0.2328	0.6640	0.1450	0.5627	0.9349	0.4654	0.7386	0.9823	**0.0333**
X_{17}	0.1785	0.2042	0.7859	0.3170	0.4197	0.1008	0.2642	0.4891	0.6132
X_{18}	0.7141	**0.0099**	0.7626	0.1667	0.9054	0.3361	0.3170	0.3361	0.8085
X_{19}	0.0867	0.9823	0.3560	0.3765	0.8885	0.4654	0.2987	0.4197	0.0744
X_{20}	0.4654	**0.0462**	0.1910	0.9054	0.7386	0.5132	0.5627	0.4197	0.1450
X_{21}	0.5378	0.8885	0.5132	0.5132	**0.0218**	0.4654	0.2181	0.5879	**0.0362**

Table 5.4. P-values for test of normality of $\widehat{f}_{z(r_P)}$

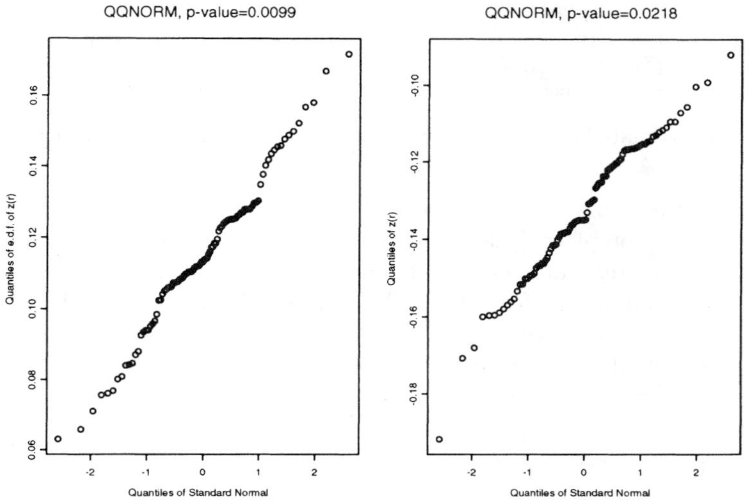

Figure 5.7. Worst approximations of the normal distribution from empirical $z(r_P)$

Although the data do not follow any normal distribution, the asymptotic works quite well as was expected.

Spearman's Rank Correlation Coefficient

However, a nonparametric alternative is offered making use of Spearman's rank correlation r_S coefficient. Kendall's τ is calculated as well. It behaves very similarly to Spearman's r_S in this data set but uses significantly more computing time; thus we only apply the correlation coefficient of Spearman. Since we observe high proportions of zeros with some variables, ties occur. Spearman's correlation coefficient is calculated following Büning and Trenkler (1978), p. 256, with

$$\begin{aligned} r_S &= \frac{(n-1)n(n+1) - 6\sum d_i^2 - 6(B_X + B_Y)}{\sqrt{((n-1)n(n+1) - 12B_X)\,((n-1)n(n+1) - 12B_Y)}}, \quad (5.6) \\ B_X &= \frac{1}{12}\sum (b_j - 1)b_j(b_j + 1) \\ B_Y &= \frac{1}{12}\sum (c_j - 1)c_j(c_j + 1) \\ d_i &= r_i - s_i \quad , i = 1, 2, \ldots, N, \end{aligned}$$

where r_i and s_i are the ranks of X and Y for $i = 1, 2, \ldots, n$. The b_j, c_j count the number of ties of the jth group of ties of X and Y, respectively. Under the hypothesis of X and Y being independent $\sqrt{n-1}r_S$ is approximately standard normally distributed. Since our number of observations

Spearman's r_S			Pearson's r_P		
$\rho = 0$	$\rho > 0$	$\rho < 0$	$\rho = 0$	$\rho > 0$	$\rho < 0$
11.04	29.28	22.68	10.36	29.96	22.68

Table 5.5. Estimated expected number of zero, positive, and negative correlations

is large, we use this approximation to test for the independence of X and Y. Using the complete data set of size $n = 5496$ we now find none of the 63 correlations regarded as zero, 40 correlations significantly positive, and 23 negative. Inferences from Kendall's τ (see Büning and Tenkler (1978), p. 264) lead exactly to the same conclusions. Thus, the nonparametric tests are not conservative, accepting more association in the data than inference from Pearson's correlation coefficient r_P proposes.

Addressing Non-Normality

To address the problem of non-normality once again, another simulation is done taking $k = 50$ random samples of size $n = 5496$ from a 30-dimensional normal distribution with estimated mean and sample covariance matrix as obtained from the complete data set. For every sample we test the hypothesis of independence for the 63 combinations of X and Y via the t-test according to (5.4) as well as the asymptotic of Spearman's r_S according to (5.6). The conclusions as to whether the null hypothesis is rejected or the correlations are regarded as positive or negative at the 5% significance level are stored for every simulation. Finally, the estimated expected number of zero, positive, and negative correlations is calculated by averaging the results from each sample and is shown in Table 5.5.

As a side effect we see how similar inferences based on Spearman's r_S and Pearson's r_P are when the data are normal and large. Comparing Table 5.5 with the results from our non-normal data in Table 5.2 we realize that we derive exactly the same conclusions from the non-normal data with the test based on the z-transformation.

Finally, remember that we are not interested in valid inference about the real world here and, moreover, the number of observations is quite large. Concerning the bivariate associations, we are mainly interested in the comparison of the complete and the imputed data and the discrepancies in inference when based on the imputed data set.

5.2.3 Conditional Independence

For the sake of simplicity we denote the covariance matrix of a vector Z of random variables following any arbitrary distribution with $var(Z) = cov(Z, Z)$. The covariance matrix $cov(Z, X)$ on the other hand refers to the covariances $\{\sigma_{ZX}\}$ only and is not necessarily symmetric.

	Y_{22}	Y_{23}	Y_{24}	Y_{25}	Y_{26}	Y_{27}	Y_{28}	Y_{29}	Y_{30}
X_{15}	**0.03**	0.32	**0.01**	**-0.02**	0.01	**0.00**	**0.00**	**0.00**	**0.01**
X_{16}	0.07	0.35	**0.00**	-0.06	**0.01**	0.04	**-0.01**	**0.00**	**-0.01**
X_{17}	0.11	1.14	0.08	-0.10	0.04	0.04	0.02	**0.00**	**0.00**
X_{18}	0.07	0.77	**0.03**	-0.06	0.02	0.04	0.01	**0.00**	**0.00**
X_{19}	0.17	1.33	0.10	-0.10	0.03	0.08	0.02	**0.00**	0.04
X_{20}	**0.01**	-0.41	**-0.03**	**-0.03**	**0.00**	**0.02**	**0.00**	**0.00**	-0.02
X_{21}	**-0.02**	-0.41	**-0.01**	**0.00**	**0.00**	**0.01**	**0.00**	**0.00**	-0.02

Table 5.6. Conditional correlation structure $\widetilde{cov}(X_i, Y_j|Z)$ of the media data

Suppose now that the covariance matrix $var(Z, X, Y)$ is partitioned as

$$
\begin{aligned}
var(Z,X,Y) &= E\left(\begin{pmatrix} Z \\ X \\ Y \end{pmatrix} - E\begin{pmatrix} Z \\ X \\ Y \end{pmatrix}\right)\left(\begin{pmatrix} Z \\ X \\ Y \end{pmatrix} - E\begin{pmatrix} Z \\ X \\ Y \end{pmatrix}\right)' \\
&= \begin{pmatrix} var(Z) & cov(Z,X) & cov(Z,Y) \\ cov(X,Z) & var(X) & cov(X,Y) \\ cov(Y,Z) & cov(Y,X) & var(Y) \end{pmatrix}; \qquad (5.7)
\end{aligned}
$$

then the conditional covariance of X and Y given $Z = z$ is

$$cov(X,Y|Z=z) = cov(X,Y) - cov(X,Z)var(Z)^{-1}cov(Z,Y), \qquad (5.8)$$

assuming the normal model for (X, Y, Z); see Whittaker (1990), p. 135. Note that the second term, $cov(X, Z)var(Z)^{-1}cov(Z, Y)$, is equal to the fusion covariance $\widetilde{cov}(X, Y)$ derived from the frequentist approach; see (2.17). According to (5.8) we calculate a conditional covariance matrix displayed in Table 5.6 using the complete data set of $n = 5496$ and the usual sample covariance matrices. Again we want to carry out a test to see how many of the conditional covariances could truly be zero.

Let us first consider the bivariate conditional independence of Y_j and X_i given the common variables Z; i.e., $(Y_j||X_i|Z)$ for $i = 15, 16, \ldots, 21$, $j = 22, 23, \ldots, 30$. Under normal theory with fixed regressors (Z, X) the sample covariance matrices $\widehat{\Sigma}_{ZX_iY_j}$ hold the sufficient statistics for the unknown parameters β and σ^2 of the $9 \times 7 = 63$ different regressions for every Y_j on (Z, X_i) for $i = 15, 16, \ldots, 21, j = 22, 23, \ldots, 30$. It is well known (see Whittaker (1990), p. 329) that the maximum likelihood estimate of the regression coefficient β of each regression is

$$\widehat{\beta}_{Y_jX_iZ} = \widehat{\Sigma}_{ZX_i}^{-1}\widehat{\Sigma}_{ZX_iY_j} \quad i = 15, 16, \ldots, 21, \ j = 22, 23, \ldots, 30. \qquad (5.9)$$

The coefficients of $\widehat{\beta}_{Y_jX_iZ}$ are known as the partial regression coefficients; see Stuart and Ord (1991), p. 1015. For each single element $\widehat{\beta}_{Y_jX_i.Z}$ therefrom we may write

$$\widehat{\beta}_{Y_jX_i.Z} = \frac{\widehat{\sigma}_{Y_jX_i|Z}}{\widehat{\sigma}_{X_i|Z}} \qquad i = 15, 16, \ldots, 21, \ j = 22, 23, \ldots, 30. \qquad (5.10)$$

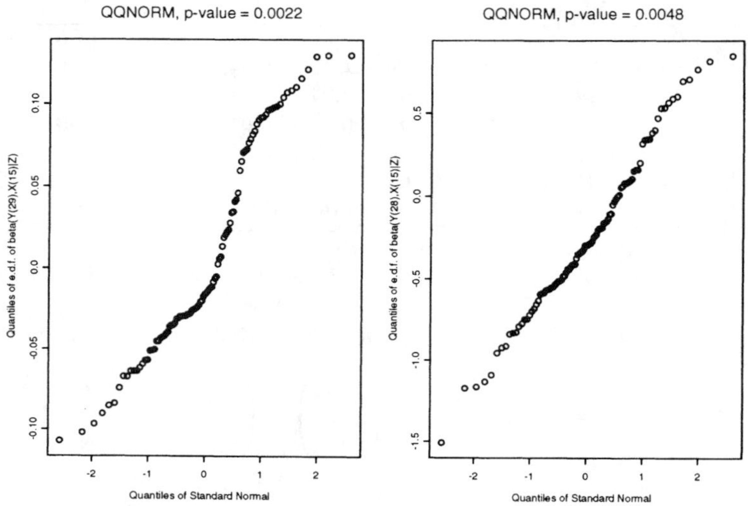

Figure 5.8. Worst approximations of the normal distribution from empirical $\widehat{\beta}_{YX.Z}$

Thus, the standard test procedure for excluding one variable from the regression is equivalent to testing the conditional independence of Y_j and X_i given the common variables $Z = z$; i.e., $\sigma_{Y_j X_i|Z} = 0$ (see Stuart and Ord (1991), p. 985). However, with these procedures the data are assumed to come from normal distributions, which precisely is not the case. On the other hand, the number of observations is very large. Whatever the underlying distribution, the least square estimates always fulfill (5.9) and (5.10) enclosing the (conditional) covariances. Nevertheless, to test the hypothesis of the conditional correlation being zero, we need to know the distribution of $\widehat{\beta}$ or at least its asymptotic behavior.

Analogous to the sections before, we perform a small simulation with $k = 100$ iterations to discuss the asymptotic behavior of the least squares estimates. Simple random samples without replacement of size $n = 2000$ are drawn from the complete data set, allowing us to compute the least squares estimates of the regression coefficients of $\beta_{Y_j X_i.Z}$ for $i = 15, 16, \ldots, 21$, $j = 22, 23, \ldots, 30$. Table 5.7 contains the p-values for the test of normality of $\widehat{f}_{\widehat{\beta}_{YX.Z}}$ applying the χ^2 goodness-of-fit test with 10 degrees of freedom again; see Section 5.2.2.

Here we would reject the hypothesis of normality for the empirical $\widehat{f}_{\widehat{\beta}_{YX.Z}}$ in only four cases at a 5% significance level, but with $p \leq 0.01$ explicitly only in two cases. These two worst approximations of the normal distribution by the empirical $\widehat{f}_{\widehat{\beta}_{YX.Z}}$ are presented as qqnorm plots in Figure 5.8.

P-values	Y_{22}	Y_{23}	Y_{24}	Y_{25}	Y_{26}	Y_{27}	Y_{28}	Y_{29}	Y_{30}
X_{15}	0.0803	0.2042	0.5879	0.5879	0.3765	0.7141	**0.0048**	**0.0022**	0.2328
X_{16}	0.7859	0.7141	0.4891	0.6891	0.1667	0.4654	0.7859	0.4197	0.3560
X_{17}	0.9054	0.5879	0.1256	0.2811	0.6891	0.6386	0.4891	0.7859	0.4654
X_{18}	0.8302	0.4654	0.7386	0.8703	0.3765	0.8508	0.4422	0.4197	0.2328
X_{19}	0.2481	0.0501	0.8703	0.3361	0.3170	**0.0426**	0.6386	0.6386	0.4422
X_{20}	0.2481	0.3765	0.9473	0.1085	0.8703	0.2042	0.2481	0.5132	0.0688
X_{21}	0.3978	0.8508	0.2328	0.6640	0.2181	**0.0168**	0.6132	0.2181	0.1350

Table 5.7. P-values for test of normality of $\widehat{f}_{\widehat{\beta}_{Y X.Z}}$

Finally, to test the hypothesis of conditional independence we need an underlying normal distribution. Whereas zero conditional correlations imply conditional independence for the multinormal, this is not true in general. Therefore, no inference about the conditional independence is drawn based on these non-normal data. We are left with testing for zero conditional correlation hereinafter as long as we do not use simulated data from a multivariate normal distribution. Applying the standard t-test of significance to the 63 regression coefficients leads to 32 relations of no conditional association (printed with bold letters in Table 5.6), 23 positive correlations, and 8 negative correlations for all the data. For $k = 50$ simulated data sets of size $n = 5496$, generated as multivariate normally distributed data with mean and covariance matrix identical to the estimated mean and covariance of the survey data again, the tests would regard on the average 32.04 from 63 correlations as being zero, 21.4 correlations as being positive, and 9.56 correlations as negative. Again there are only the slightest differences of inference (see Table 5.6) when based on the non-normal data.

Moreover, we compute the deviance against conditional independence of the variables X and Y given the common Z variables using the complete data with

$$
\begin{aligned}
dev(Y||X|Z) &= -n \ln \left(\frac{\det(\widehat{\Sigma}_{XYZ})\det(\widehat{\Sigma}_Z)}{\det(\widehat{\Sigma}_{XZ})\det(\widehat{\Sigma}_{YZ})} \right) \qquad (5.11) \\
&= -5496 \ln \left(\frac{9.4955 \cdot 10^{27} \cdot 1.6449 \cdot 10^{27}}{5.0474 \cdot 10^{15} \cdot 3.2301 \cdot 10^{39}} \right) = 235.78\,,;
\end{aligned}
$$

see Whittaker (1990), p. 331. Under the hypothesis of conditional independence this deviance is asymptotically χ^2 distributed with 63 degrees of freedom. Since $\chi^2_{0.95}(63) = 82.53$ the hypothesis of X and Y being independent given $Z = z$ will be rejected for the complete data set. Note that from the 50 randomly generated data sets we get a mean value of 300.00 for the deviance.

Since we do not want to describe the relationship of the variables in a meaningful way, we use the original data set and the imputed data throughout nearly the whole study without laying emphasis on transformations to do the comparisons after the match. Anyway, the structure and volume of our survey data enable us to discuss three levels of validity of the statistical matching techniques as developed in Chapter 2: first, preserving the univariate structure of the data, second, recovering the correlation structure, and third, with some restrictions, estimating the joint distribution of the variables not jointly observed.

5.3 Simulation Study Using Generated Data

When analyzing the complete survey data set, we realized that there is a large amount of bivariate conditional "independence" or, at least, no correlation of X and Y given $Z = z$ already. Since half of the combinations of X and Y are obviously conditionally uncorrelated, even the traditional matching techniques are expected to be quite efficient. To investigate the matching techniques under circumstances with only a few variables X and Y being conditionally independent given $Z = z$, we multiply the estimated conditional covariance by the factor c and calculate the "new" unconditional association $cov(X, Y)^\star$ according to (5.8) with

$$cov(X,Y)^\star = c \cdot \widehat{cov}(X, Y|Z = z) + \widehat{cov}(X, Z)\widehat{var}(Z)^{-1}\widehat{cov}(Z, Y) \quad (5.12)$$

with $c = 6$. The ordinary sample covariances are taken as estimates. The original sample covariance $\widehat{cov}(X, Y)$ is replaced by $cov(X, Y)^\star$ giving a new complete covariance matrix $var(Z, X, Y)^\star$,

$$var(Z, X, Y)^\star = \begin{pmatrix} \widehat{var}(Z) & \widehat{cov}(Z, X) & \widehat{cov}(Z, Y) \\ cov(Z, X) & \widehat{var}(X) & cov(X, Y)^\star \\ \widehat{cov}(Z, Y) & \widehat{cov}(X, Y)^\star & \widehat{var}(Y) \end{pmatrix}, \quad (5.13)$$

which still is positive definite. Notice that if the factor c is set to $c = 7$ a negative determinant is produced, so we consider $c = 6$ a good choice.

Now samples of size $n = 5496$ are drawn randomly $k = 50$ times from a 30-dimensional multivariate normal distribution with mean $\widehat{\mu}$ and covariance $var(Z, X, Y)^\star$. On the average only 6.46 conditional associations of X and Y given Z are regarded as zero at a significance level of 5%. 35.80 conditional associations are obviously positive and 20.74 are negative. The deviance for the two blocks of variables X and Y given $Z = z$ has a value of 24600.69 on the average.

A proven method for performing simulation studies with generated data, e.g., would be to draw, say $k = 1000$ random samples of size $n = 5496$ from this multivariate normal distribution; see among many others Schafer (1997). Patterns of missingness are usually imposed on these random samples. Then imputations are created and point as well as interval estimates for the underlying population parameters are calculated to evaluate the validity of the imputation techniques. We do not follow this approach hereinafter because our objective is a little bit different. We want to investigate the extent to which the results of our simulation study using the survey data are due to departures from the normal model, and how efficient the techniques are if conditional independence scarcely exists. Since the generated data come from a normal distribution, we no longer have to distinguish between conditional independence and conditional "noncorrelation" now. Therefore we take one generated data set which seems to be quite representative for this parameter setting and rerun exactly the same investigations as we do when using the survey data. All analyses in the following eval-

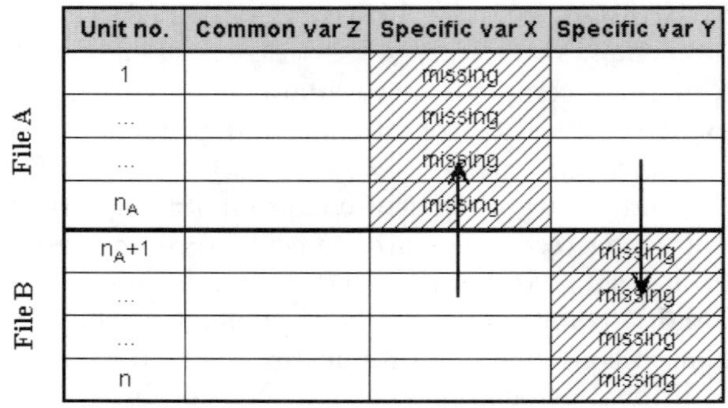

Figure 5.9. File concatenation in general

uation study are done twice: first the real survey data and second the simulated data are used for it.

The data set we choose shows 7 combinations of X and Y being conditionally independent, 35 regarded as related positive, and 21 as negative, respectively. The deviance against conditional independence is calculated with 24852.50.

5.4 Design of the Evaluation Study

We divide the data into two parts, whereas we systematically sample every second individual for the file B. Numbered from one to n, again a data set known as file concatenation (see Rubin (1986)) is created according to Figure 5.9.

After having imputed the missing data the artificial data set can be compared with the original data set for every matching direction; i.e., we compare the imputation of the X variables in file A as well as the imputation of the Y variables in file B. This is equivalent to adding the TV viewing information to the purchasing behavior panel as well as adding the purchasing information to the television measurement panel, as is done in practice by the GfK and others; see Chapter 3.

Notice that we again do not use the original survey data set as the underlying population with the ultimate ambition to provide efficient point and interval estimates for its parameters. Generating k (e.g., $k = 1000$) random samples of size $n = 500$ or 1000 from such a population, imposing patterns of missingness, creating imputations, calculating point and interval estimates, and counting the coverage[6] is quite familiar for evaluating multiple imputation techniques; see e.g. Schafer (1997), p. 216.

However, we follow the simulation studies conducted by Kamakura and Wedel (1997) as well as Raghunathani and Grizzle (1995), where inferences obtained using the complete data and the imputed (i.e., the matched data) are compared. This comparison may suggest how much is lost by basing inference on matched samples rather than on single source samples, at least with this type of television and purchasing behavior data and the proposed statistical matching techniques. Moreover, this comparison aims at evaluating whether the proposed alternative approaches may yield "proper" multiple imputations in the sense defined by Rubin (1987).

5.4.1 Techniques Under Investigation

Throughout the evaluation the following settings are used.

- The data set is cut into halves, delivering file A with (Z_A, Y_A) observed and X_A missing, as well as file B with (Z_B, X_B) observed and Y_B missing. We compare the effect of imputing X_A from X_B and Y_B from Y_A.
- First we execute the imputations without use of any auxiliary file. Then auxiliary files of size $n = 300$ and $n = 600$ are drawn from the complete data set before dividing them into two parts.
- Three different statistical matching techniques are implemented for the study, i.e.,
  - the new noniterative multivariate model NIBAS,
  - data augmentation NORM using a ridge prior, and
  - the flexible imputation using MICE.
- These alternative techniques are applied also to a simulated data set with only a few variables X and Y being conditionally independent given $Z = z$.

Altogether we get $2 \cdot 3 \cdot 3 \cdot 2 = 36$ different settings to which we restrict our research.

5.4.2 Measuring the Goodness-of-Fit

To start with a measure of the fit for the distributions before and after the statistical matching, we look at the univariate distributions of the variables X_{16}, X_{17}, and X_{18} as well as the variables Y_{22}, Y_{23}, and Y_{24} by means of a

[6]The coverage is the number of interval estimates, e.g., out of 1000 simulations that cover the true population value.

histogram plot as done before and by a qqplot, i.e., plotting the quantiles of the c.d.f of the imputed variable in file A or file B against the c.d.f of the original variable in the same file. Moreover, the structure of each imputed variable is shown analogous to Table 5.1. Since this is a visual comparison, tests are performed immediately afterwards.

Most of the following tests are applied on the original and imputed data set without transforming the data to achieve at least univariate normality. Only when standard two-sample tests are performed, do we use transformations. We have already discussed the validity of some tests in this chapter although the raw data are not normally distributed. Finally, we are not interested in the scientific meaningful inference concerning the real data, but we want to compare whether the imputed data provide the same conclusions as the original data. We have argued in Section 4.3.3 how difficult it is to show whether an imputation procedure is a good procedure or in terms of Rubin (1987) leads to proper multiple imputations. Roughly speaking, when the data are regarded as fixed and the missingness mechanism is repeated, then with an increasing number of imputations proper MI should yield the same inference that would be drawn if the data had been completely observed. To discover the similarities in inference based on the original complete data and the imputed data such tests are applied which in principle allow us to compare two samples. Moreover, the tests we choose concern different levels of validity according to our terminology introduced in Section 2.5.

Systematically we discuss the following topics and levels of validity for every matching technique we perform:

(1) Comparison of marginal distributions (concerns the fourth level),

(2) bivariate associations (concerns the third level),

(3) conditional independence (concerns the third level),

(4) cross-tabulation fit (concerns the second level), and

(5) linear regression inference (concerns the third level).

The implemented test procedures are described in detail in the following sections.

Comparison of Marginal Distributions

First of all, we test the null hypothesis that the proportions p^{org} and p^{imp} of zeros are equal for the imputed values of X (and Y) and the original values with the standard t-test according to

$$\begin{aligned} t &= \frac{\widehat{p}^{org} - \widehat{p}^{imp}}{\sqrt{\frac{(n^{org}-1)\widehat{p}^{org}(1-\widehat{p}^{org})+(n^{imp}-1)\widehat{p}^{imp}(1-\widehat{p}^{imp})}{n^{org}+n^{imp}-2}}} \sqrt{\frac{n^{org} n^{imp}}{n^{org} + n^{imp}}} \\ &\sim t(n^{org} + n^{imp} - 2), \end{aligned} \qquad (5.14)$$

with $\hat{p}^{org}$ and $\hat{p}^{imp}$ denoting the sample proportion of zeros for the original and imputed data, for X and Y, respectively.

A two sample Kolmogorov–Smirnov goodness-of-fit test is performed, which tests the equality of two distributions for variables from X^{org} and X^{imp} (and from Y^{org} and Y^{imp}) using the S-PLUS routine ks.gof. The test statistic depends on the greatest absolute vertical distance between the empirical distribution functions of X^{org} and X^{imp} (and Y^{org} and Y^{imp}); for details see Büning and Trenkler (1978), p. 133. The Kolmogorov–Smirnov test is known to be rather conservative, especially when ties occur.

Thus, we remove the observed zero values from each variable X and Y getting the subsets X_+ and Y_+ and perform a nonparametric test, whether the median of the distribution of $D = X_+^{org} - X_+^{imp}$, respectively, $= Y_+^{org} - Y_+^{imp}$ is zero within the remaining n observations. Therefore the Wilcoxon signed rank test for paired sample data is applied via the S-PLUS routine wilcox.test. The test statistic W is the rank sum of those $|D|$ where $D > 0$ holds. Since the number of observations is large and ties still may occur, the standard normal approximation is used for $(W - n(n-1)/4)/\sqrt{n(n+1)(2n+1)/24}$; see Büning and Trenkler (1978), p. 198.

Finally, the n remaining values are transformed on a log scale to achieve univariate normality as described in Section 5.2.1. Then the standard two-sample paired t-test is conducted with $D = V^{org} - V^{imp}$ and the test statistic

$$t = \frac{\overline{D}}{\sqrt{\widehat{var}(D)}} \sqrt{n} \sim t(n-1)\,. \tag{5.15}$$

Under the null hypothesis that the population mean of the difference D is equal to zero, t follows a t-distribution with $n - 1$ degrees of freedom assuming normality of the differences D. Since we use the transformed data, this assumption may not be violated too much.

Bivariate Associations

The proposed matching techniques can be assessed further by comparing the results of the tests of association from the imputed data with those obtained from the original, i.e., the complete data.

First the nonparametric alternative is calculated with Spearman's rank correlation coefficient making use of the S-PLUS routine cor.test. The statistic was already described in Section 5.2.2 with (5.6). According to Spearman's rank correlation coefficient in file A we regard 4 correlations as being zero, 36 correlations as positive, and 23 correlations as negative, whereas in file B no correlations are regarded as zero, 40 are obviously positive, and 23 are negative.

Encouraged by the results of the asymptotic behavior of the z-transformation according to (5.5) presented in Section 5.2.2, we further-

Imputed data set

Complete data set / Bivariate ass.	Uncorrelated	Pos. association	Neg. association	Total complete	
Uncorrelated	Correct specified	False positive	False positive	File A: 15	File B: 14
Pos. association	False negative	Correct specified	Contradiction	File A: 25	File B: 27
Neg. association	False negative	Contradiction	Correct specified	File A: 23	File B: 22
Total imputed				63	63

Figure 5.10. Counting bivariate associations

more use this test statistic and its approximation with

$$z(r_p)\sqrt{n-3} \quad \sim (\text{appr.})\, N(0,1), \quad r_P = \widehat{\rho}(X_i, Y_j), \tag{5.16}$$

for $i = 15, 16, \ldots, 21$, $j = 22, 23, \ldots, 30$ under the hypothesis of $\rho(X_i, Y_j) = 0$, i.e., X_i and Y_j being uncorrelated for $i = 15, 16, \ldots, 21$, $j = 22, 23, \ldots, 30$.

We take the original and the imputed data without any transformations. The conclusions at the 5% significance level are counted and compared with the conclusions based on the original data set as Figure 5.10 shows.

The number of reported associations differs only a bit between the two files.

Not to bother the reader with endless lists of those figures, we condense the information of Figure 5.10 by counting and presenting only the correct specified associations, the false positives, i.e., accepting association in the imputed data when there was none in the complete data, and false negatives, i.e., accepting no correlation when there was association. Sometimes even an oppositional conclusion is reported.

Conditional Independence

As described in Section 5.2.1, the test of bivariate conditional independence of $(Y_j||X_i|Z)$ for $i = 15, 16, \ldots, 21$, $j = 22, 23, \ldots, 30$ is equivalent to testing whether the corresponding regression coefficient is zero. Again we find the asymptotic behavior of $\widehat{\beta}_{YX.Z}$ as discussed in Section 5.2.3 satisfying enough to use the standard test statistic and its asymptotic

$$\frac{\widehat{\beta}_{Y_j X_i . Z}}{\sqrt{\widehat{var}(\widehat{\beta}_{Y_j X_i . Z})}} \sim (asy)\, N(0,1) \tag{5.17}$$

under the hypothesis of $\beta_{Y_j X_i . Z} = 0$, i.e., X_i and Y_j being (conditionally) uncorrelated given $Z = z$ for $i = 15, 16, \ldots, 21$, $j = 22, 23, \ldots, 30$. Then the correct specified correlations are counted as well as the false positives, the false negatives, and the contradictions as displayed in Figure 5.11.

In file B we realize a larger amount (47) of conditional noncorrelations of X and Y given $Z = z$ within the complete data than in file A (37). Notice that this is reversed when the auxiliary file of size 300 is taken from

Imputed data set

Complete data set (rows)

Conditional ass.	Uncorrelated	Pos. association	Neg. association	Total complete	
Uncorrelated	Correct specified	False positive	False positive	File A: 37	File B: 47
Pos. association	False negative	Correct specified	Contradiction	File A: 20	File B: 12
Neg. association	False negative	Contradiction	Correct specified	File A: 6	File B: 4
Total imputed				63	63

Figure 5.11. Counting conditional associations

the data. For the remaining $n - 300 = 5196$ observations we found for the 2598 units of file A (B) 50 (36) of 63 correlations regarded as zero. When discussing the use of the auxiliary files in later sections, we take this into account.

Moreover, we present the deviance against conditional independence of X and Y given $Z = z$ in the complete data as well as the imputed data set according to (5.11).

Cross-Tabulation Fit

To assess the degree of fit between the complete and our imputed distributions, we process all 63 possible pairs of the 7 X variables and the 9 Y variables and compare the cell frequencies from the imputed data set with those from the complete data set. Since our variables are semicontinuous we cut each variable into 5 categories and calculate the frequencies $n_{ij}, i, j = 1, 2, \ldots, 5$ of $5 \times 5 = 25$ cells for each of the 63 cross-tabulations. According to the evaluation of Kamakura and Wedel (1997) we compute a correlation coefficient between the cell frequencies of the complete data and the imputed data.

The correlation coefficient r between n_{ijk}^{org} and n_{ijk}^{imp}, $i, j = 1, 2, \ldots, 5$, $k = 1, 2, \ldots, 63$ is calculated comparing these frequencies for all 63 tables of the complete data and the imputed data according to

$$r = \frac{\sum_{k=1}^{63}\sum_{i=1}^{5}\sum_{j=1}^{5}(n_{ijk}^{org} - \overline{n}^{org})(n_{ijk}^{imp} - \overline{n}^{imp})}{\sqrt{\sum_{k=1}^{63}\sum_{i=1}^{5}\sum_{j=1}^{5}(n_{ijk}^{org} - \overline{n}^{org})^2 \sum_{k=1}^{63}\sum_{i=1}^{5}\sum_{j=1}^{5}(n_{ijk}^{imp} - \overline{n}^{imp})^2}}. \tag{5.18}$$

Figure 5.12 illustrates this calculation procedure.

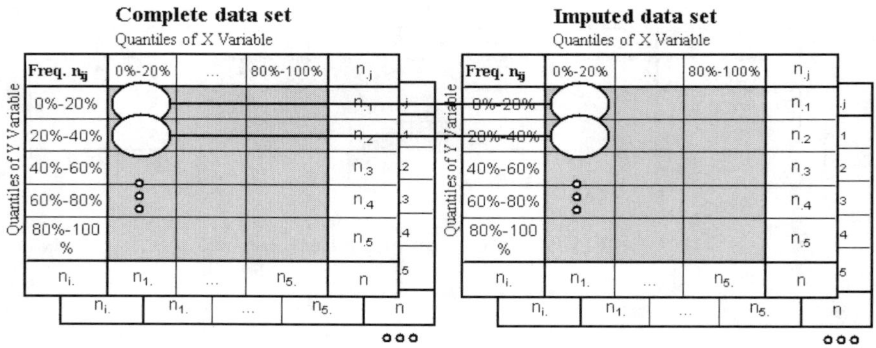

Figure 5.12. Actual and imputed cross-tabulations of X and Y

The higher r is the better the "unknown" frequencies are reproduced by the imputation.

However, Kamakura and Wedel (1997) pointed out that the correlation coefficient calculated on the basis of the cell frequencies might overestimate the power of the matching techniques to reproduce the true associations of the X and Y variables. They argue that parts of the correlation coefficient could be due to the model's ability to fit the row and column margins of the variables being cross-tabulated. By calculating the correlation coefficient r_{KW} using the deviations of the cell frequencies from the expected frequencies under the hypothesis of independence between rows and columns, we find a lower r_{KW} in some cases and higher r_{KW} in others. Our variables are continuous and, thus, the margins of X and Y are not fixed a priori, therefore we do not follow the argumentation of Kamakura and Wedel (1997). Just to make our results comparable to their work in some sense, we compute and present the correlation coefficient r_{KW} using the differences $n_{ijk}^{org} - n_{i.k}^{org} n_{.jk}^{org}/n$ and $n_{ijk}^{imp} - n_{i.k}^{imp} n_{.jk}^{imp}/n$, $i, j = 1, 2, \ldots, 5$, $k = 1, 2, \ldots, 63$ instead of n_{ijk}^{org} and n_{ijk}^{imp} in (5.18).

Moreover, an ordinary χ^2 test of homogeneity (see Mood et al. (1974), p. 449) is performed to check if these bivariate distributions from the original and the imputed data sets can be considered the same at a significance level of 5%. Therefore we compare the 25 cell frequencies n_{ijk}^{org} with the 25 cell frequencies n_{ijk}^{imp} and calculate the χ^2 test statistic for every cross-tabulation $k = 1, 2, \ldots, 63$; see Figure 5.13.

Finally we count the number of times out of 63 the null hypothesis is not rejected; i.e., we may consider the two cross-tabulated samples as being drawn from the same population.

Freq.	$Cell_{11}$	$Cell_{12}$	...	$Cell_{54}$	$Cell_{55}$	Sum
n_{ij}^{org}						n^{org}
n_{ij}^{imp}						n^{imp}
Sum	n_{11}	n_{12}	...	n_{54}	n_{55}	$n=n^{org}+n^{imp}$

Sum	n_{11}	n_{12}	...	n_{54}	n_{55}	$n=n^{org}+n^{imp}$

ooo

Figure 5.13. Test of homogeneity based on 25 cells

Imputed data set

Complete data set: Effect	No effect	Pos. effect	Neg. effect	Total complete	
No effect	Correct specified	False positive	False positive	File A: 141	File B: 142
Pos. effect	False negative	Correct specified	Contradiction	File A: 28	File B: 30
Neg. effect	False negative	Contradiction	Correct specified	File A: 29	File B: 26
Total imputed				198	198

Figure 5.14. Counting linear regression inferences

Linear Regression Inference

Along with the univariate and the bivariate analyses we perform some regression analyses with these data. Since it might be of interest to estimate the influence of viewing a particular TV channel on purchasing behavior, we calculate all possible multivariate linear regressions from each Y_j variable on all (imputed) variables X and the common variables Z in file A as well as in file B, where the Y values are the imputed ones. The estimated model therefore is the ordinary linear one with

$$Y_j = \beta_{0j} + \beta_{1j} Z_1 + \ldots + \beta_{21j} X_{21} + U_j, \quad U_j \sim N_{n_A}(0, \sigma^2 I), \tag{5.19}$$

for $j = 22, 23, \ldots, 30$. Thus we get $9 \times 22 = 198$ conclusions (including the constant) about the significance of the regression coefficients to compare using the original as well as the imputed data for files A and B, respectively. Again we count the number of correct specifications, of false positives, false negatives, and contradictions, if there are any, according to Figure 5.14.

Similarly to the bivariate associations, inferences drawn separately from the two files do not differ much.

5.4.3 Overview

For the simulation study using the generated data all the tests according to Section 5.4.2 may be applied without further discussion because the data are drawn from a multivariate normal distribution.

The following list provides an overview of different estimates $\widehat{\theta}$ and test statistics that are used hereinafter. Notice that the data are transformed by splitting them into a binomial and a continuous part as described in Section 5.2.1. The continuous part is transformed to achieve normality before the imputations are generated by NIBAS or NORM. After the imputations are performed the data are transformed back to their original scale. Excluding the standard two-sample paired t-test all the tests are based on this back-transformed data. In Sections 5.2.2 and 5.2.3 we discussed the validity of the tests that are applied although the raw data are not normally distributed. Remember that the focus of the study is not to derive meaningful inference about the underlying population but to show whether inference from the original data will differ from inference based on the imputed data.

When we apply multiple imputation techniques for fusing the data m times, then inference from the imputed data should be based on the corresponding multiple imputation estimate $\widehat{\theta}_{MI}$, which is simply the mean of the m single imputation estimates $\widehat{\theta}^{(k)}$. Note that the multiple imputation principle demands averaging the statistic that would be used to estimate θ if the complete data were available.[7] Tests are based on the Student's t-approximation $(\widehat{\theta}_{MI} - \theta^0)/\sqrt{T} \sim t$ then. To obtain the test-statistic for the MI estimate we must calculate the estimated total variance $T = (1 + m^{-1})B + W$ from the between-imputation variance $B = (1/(m-1)) \sum(\widehat{\theta}^{(k)} - \widehat{\theta}_{MI})^2$ as well as the within-imputation variance $W = (1/m) \sum \widehat{var}(\widehat{\theta})^{(k)}$ as described in detail in Section 4.3. To clarify this calculation, the list below also accounts for the variance estimate on which each calculation of the within-imputation variance according to (4.24) is based. Remember that for performing the multiple imputation test procedures we have to calculate the single imputation estimates and their single imputation variances which is quite different from averaging the single imputation test-statistics themselves.

(1) Comparison of marginal distributions:
According to the multiple imputation theory proposed by Rubin (1987) rules are given to combine parameter estimates to achieve the multiple imputation estimate and derive valid inference therefrom. Li et al. (1991) also provide strategies for computing significance levels from multiply imputed data sets concerning more general tests but these strategies are not addressed to compare distributions. Therefore, we find it more informative to evaluate each single imputed data set and compare it with the original data by using standard test procedures than to produce multiple imputation estimates, e.g., of the proportion of zeros.

[7] Inferential questions that cannot be cast in terms of a one-dimensional parameter, such as, e.g., many goodness-of-fit tests, may be handled through some multivariate generalizations but are quite unfamiliar yet; see Li et al. (1991) or Meng and Rubin (1992).

Hence we calculate the p-values for each test, i.e., equal proportions, Kolmogorov–Smirnov, Wilcoxon, and paired t-test, for each imputed data set and count the number of times out of m the null hypothesis is not rejected.

(2) Bivariate associations:
The multiple imputation estimate $(1/m)\sum r_S^{(k)}$ is calculated and $1/(n-1)$ is used for the estimated variance $\widehat{var}(r_S)$. $(1/m)\sum z(r_P)^{(k)}$ is computed with $1/(n-3)$ for its estimated variance $\widehat{var}(z(r_P))$.

(3) Conditional independence:
The multiple imputation estimate for each $(1/m)\sum \widehat{\beta}^{(k)}_{Y_j X_i . Z}$ is computed and $\widehat{var}(\widehat{\beta}_{Y_j X_i . Z})$ is taken from each regression and each imputed data set. The "independence" of the blocks X and Y given $Z = z$ is tested with $dev(Y||X|Z)$ against the quantile of the $\chi^2(63)$ distribution. For the deviance we display the mean $dev(Y||X|Z)$ and count the number of times out of m the hypothesis of conditional independence is rejected.

(4) Cross-tabulation fit:
From the cross-tabulation of the 63 X and Y combinations that are categorized into $5 \cdot 5 = 25$ cells each, the correlation coefficient is calculated for every imputed data set. For the worst and the best cross-tabulation fit out of m correlation coefficients a scatter plot is displayed comparing the imputed and the original frequencies. Then the mean correlation coefficients $(1/m)\sum r^{(k)}$ and $(1/m)\sum r^{(k)}_{KW}$ are given.
Moreover, for every two pairs of the X and Y variables we test on homogeneity, i.e., whether the bivariate categorized distributions may come from the same underlying population, and count the number of those not rejecting the null hypothesis such as $\sum_{i=1}^{63} I_i$ with $I_i = 1$, if $\chi_i^2 \leq \chi_{24}^2$ and $I_i = 0$, if $\chi_i^2 > \chi_{24}^2$ on a significance level of 95%. From these χ^2 tests the minimum and the maximum number of counts from the m imputed data are recorded.

(5) Linear regression inference:
The multiple imputation estimate $(1/m)\sum \widehat{\beta}^{(k)}_{Y_j X_Z}$ is computed and $\widehat{var}(\widehat{\beta}_{Y_j X Z})$ is used from each regression and each imputed data set.

The results of the alternative imputation techniques and the test procedures proposed are presented and discussed as briefly as possible in the following sections. The main conclusions concerning the validity of a matching technique and its applicability to real data are summarized in a final paragraph at each section's end.

Name	Procedure	Iterative		Multivariate		Software	Main literature
		Yes	No	Yes	No		
NORM 2.02	Data augmentation	X		X		Windows	Schafer (1997)
MICE V1.0	Chained equations	X			X	S-PLUS	Van Buuren et al. (1999)
NIBAS V0.9	Multivariate linear regression		X	X		S-PLUS	Box and Tiao (1973)

Figure 5.15. Bayesian routines used in the study

5.5 Results Due to Alternative Approaches

For the multiple imputation procedures, we take $m = 5$ imputations. Clearly, this is quite a small number if the fraction of missing information is large. But with 50% or less missing information as is the case herein, a multiple imputation estimate based on $m = 5$ imputations has a standard deviation that is only about 5% wider than one based on the theoretically $m = \infty$ imputations due to its relative efficiency $(\sqrt{1 + 0.5/5})^{-1} = 0.9535$; for details see Section 4.3.

After imputing $m = 5$ times, five sets of the usual complete-data point estimates and their variance estimates were calculated from the data. The results were combined in the usual way for multiple imputation inference; see Sections 4.3 and 5.4.3.

Multiple imputation is constructed to yield valid inference for estimates in the case of incomplete data taking into account the uncertainty of the imputation. Now our focus is less the calculation of a good estimate for some unknown population quantities, but rather the good reproduction of joint distributions never jointly observed by the matching techniques. Therefore we combine the m estimates to ensure for the uncertainty of the missingness mechanism and base inference from the imputed data on the multiple imputation estimates. Moreover, we explore whether we may stay with one single imputation even if we apply classical multiple imputation procedures and to what extent this affects the inference. Our argument is somewhat affected by the needs of market research institutes to present one single complete sample to their customers. Their clients often perform individual analysis long after the imputations have been done, neither willing nor used to combine estimates from m data sets. Although the techniques of multiple imputation have been developed for more than 20 years now (see Rubin (1996)), acceptance is only slowly growing among practitioners.

Recently more software packages or routines have been developed realizing the Bayesian imputation routines for any missing data problem. An overview of the majority of currently available programs is given in Section 6.1.2. In our simulation study hereinafter we restrict the investigation to the three imputation procedures NIBAS, NORM, and MICE already described in Chapter 4. Figure 5.15 highlights some of their features.

It is to be expected that the availability of such software packages and routines will help spread those procedures in particular and multiple imputation in general among practitioners. These packages can be downloaded for free from the Internet; NORM 2.02 is available at http://www.stat.psu.edu/~jls/misoftwa.html, MICE V1.0 is available at http://www.multiple-imputation.com and, finally, the noniterative multivariate Bayesian regression model is implemented as the S-PLUS routine called NIBAS and provided as source code in the appendix.

Although we have introduced the regression imputation with random residual in Section 4.4, we do not consider this procedure any longer. RIEPS is not at all a proper imputation method as we have shown in the simulation study at the end of Chapter 4. RIEPS serves as an introductory basis for the new noniterative multivariate imputation procedure NIBAS which is discussed in detail in Section 4.5 and hereinafter.

Notice that MICE does not allow the use of prior information. Only a few of the currently available software packages allow the efficient use of prior information; see, e.g., figure 6.1 in the next chapter. MICE is chosen because the regression approach based on Rubin (1987), pp. 166–167, is implemented in some other multiple imputation routines in similar ways. Moreover, MICE is available for free and has an easy to use S-PLUS library. Thus, we regard MICE as representative of other multiple imputation techniques based on uninformative prior information. The simulation study at the end of Chapter 4 shows clearly that MICE establishes conditional independence in this small example and with valid model assumptions. The focus of this investigation now is the comparison of our three approaches when real data are used and the model assumptions are not well fulfilled. As a matter of fact we have to set the prior of the conditional association to zero when NIBAS is used and whenever no auxiliary information is used. For NORM a ridge prior is applied to stabilize computations. The parameter settings are described in detail at the beginning of each section.

5.5.1 Noniterative Multivariate Imputation Procedure

Now the new non-iterative routine NIBAS as described with Algorithm 2 is applied.[8] Without an auxiliary file, the conditional association of X and Y given $Z = z$ is set to zero with $cov(X,Y|Z = z) = 0$; otherwise these covariances are calculated according to (5.8) using the sample covariance matrices from auxiliary file a; i.e., $\widehat{cov}(X,Y|Z=z)_a = \widehat{cov}(X,Y)_a - \widehat{cov}(X,Z)_a \widehat{var}(Z)_a^{-1} \widehat{cov}(Z,Y)_a$ based on 300, respectively, 600 observations.

First of all we compare the marginal distributions of the imputed variables with their original distribution when no auxiliary file is used. Figure 5.16 shows the histogram plots for the imputed variables X_{15}, X_{16}, and X_{17} and their qqplots comparing the imputed data with the original data

[8]See S-PLUS function NIBAS in the appendix.

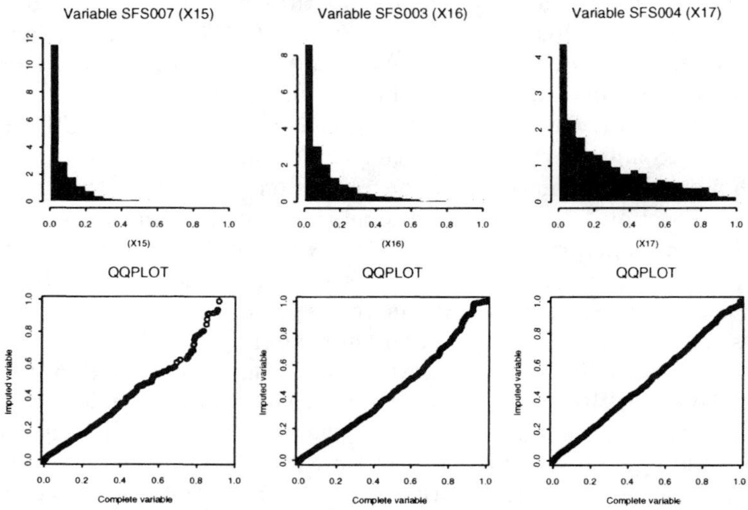

Figure 5.16. NIBAS: histogram and qqplot of imputed variables X_{15}, X_{16}, and X_{17}

in file A exemplifying for the first of the multiple imputations. The structure of the imputed data of X with $n_A = 2748$ in file A is listed in Table 5.8. Notice that no zero values for the variables X_{20} and X_{21} are generated. If Y is imputed in file B the visual comparison gives a similar picture to that shown in Figure 5.17. Some rather large y-values are generated. The structure of the imputed data of Y in file B with $n_B = 2748$ is listed in Table 5.9.

Obviously the marginal distributions do not differ too much especially for the X variables. In the following section tests are performed to see whether the estimated parameters differ significantly.

Variable	% Zeros		Mean		Standard error	
X_i	p_i^{org}	$\frac{1}{5}\sum p_i^{(k)imp}$	$\overline{X}_i^{org}$	$\frac{1}{5}\sum \overline{X}_i^{(k)imp}$	S_i^{org}	$\frac{1}{5}\sum S_i^{(k)imp}$
X_{15}	0.31	0.34	0.09	0.08	0.14	0.13
X_{16}	0.20	0.23	0.17	0.15	0.21	0.19
X_{17}	0.12	0.13	0.28	0.27	0.26	0.26
X_{18}	0.14	0.16	0.18	0.17	0.19	0.19
X_{19}	0.05	0.03	0.42	0.42	0.28	0.29
X_{20}	0.02	0.00	0.43	0.42	0.25	0.25
X_{21}	0.01	0.00	0.49	0.49	0.25	0.26

Table 5.8. NIBAS: marginal distributions of imputed variables X_{15}, X_{16}, and X_{17}

Variable	% Zeros		Mean		Standard error	
Y_i	p_i^{org}	$\frac{1}{5}\sum p_i^{(k)imp}$	$\overline{Y}_i^{org}$	$\frac{1}{5}\sum \overline{Y}_i^{(k)imp}$	S_i^{org}	$\frac{1}{5}\sum S_i^{(k)imp}$
Y_{22}	0.07	0.05	10.72	12.06	10.83	17.67
Y_{23}	0.22	0.27	33.66	29.09	77.78	96.67
Y_{24}	0.27	0.31	6.22	5.21	9.88	9.57
Y_{25}	0.05	0.03	8.22	9.74	8.56	16.70
Y_{26}	0.37	0.40	1.38	1.06	2.48	1.73
Y_{27}	0.09	0.08	8.95	9.97	8.14	13.73
Y_{28}	0.38	0.41	1.80	1.42	3.40	2.90
Y_{29}	0.32	0.36	0.04	0.01	0.27	0.02
Y_{30}	0.32	0.36	1.55	1.10	3.83	2.65

Table 5.9. NIBAS: marginal distributions of imputed variables Y_{22}, Y_{23}, and Y_{24}

Goodness-of-Fit

We first discuss the results of the imputation when no auxiliary files are available denoted by the column headed "No a" in the following tables. The conditional association $\rho_{XY|Z}$ has to be chosen a priori and is set to zero. Then we incorporate information from auxiliary files of size $n = 300$ and $n = 600$ by estimating $\rho_{XY|Z}$ from these auxiliary data. The results are reported in the following tables denoted by columns headed "a_{300}" and "a_{600}".

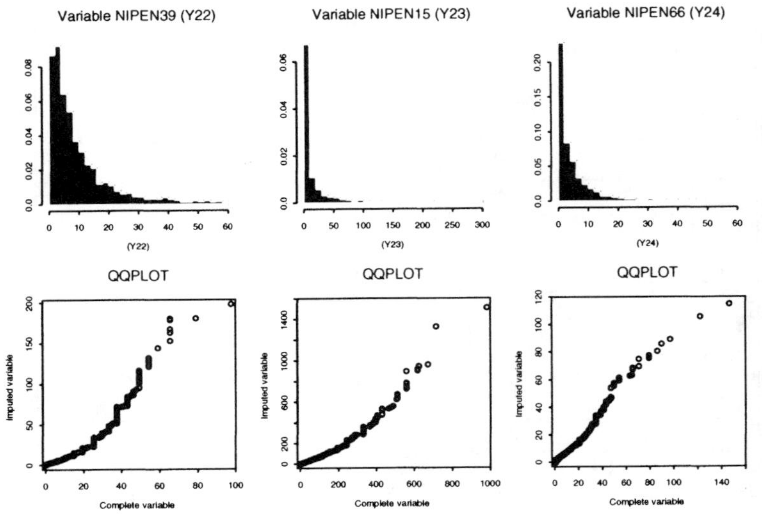

Figure 5.17. NIBAS: histogram and qqplot of imputed variables Y_{22}, Y_{23}, and Y_{24}

File	Equal % zeros			KS			Wilcoxon			Paired t		
A	No a	a_{300}	a_{600}	No a	a_{300}	a_{600}	No a	a_{300}	a_{600}	No a	a_{300}	a_{600}
X_{15}	0	0	0	0	0	0	3	2	4	4	5	5
X_{16}	0	2	3	0	0	3	4	4	5	4	5	5
X_{17}	4	5	3	4	2	3	4	5	4	5	5	4
X_{18}	3	4	5	0	0	1	1	4	4	5	5	4
X_{19}	1	0	0	5	0	1	4	3	4	5	5	5
X_{20}	–	–	–	2	3	3	2	4	3	4	5	2
X_{21}	–	–	–	2	2	3	4	2	5	4	2	4
$\sum$	8	11	11	13	9	14	22	24	29	31	32	29

File	Equal % zeros			KS			Wilcoxon			Paired t		
B	No a	a_{300}	a_{600}	No a	a_{300}	a_{600}	No a	a_{300}	a_{600}	No a	a_{300}	a_{600}
Y_{22}	1	0	0	0	0	0	3	1	0	5	4	2
Y_{23}	0	3	0	0	0	0	0	2	2	4	5	5
Y_{24}	0	0	1	0	0	0	1	5	1	5	0	4
Y_{25}	0	0	0	0	0	0	0	3	4	3	4	5
Y_{26}	3	3	1	0	0	0	0	4	3	4	4	5
Y_{27}	5	5	0	0	0	0	4	0	2	4	5	4
Y_{28}	1	4	5	0	0	0	3	2	3	3	5	4
Y_{29}	0	1	0	0	0	0	0	1	0	5	4	4
Y_{30}	1	2	0	0	0	0	5	3	4	5	5	5
$\sum$	11	18	7	0	0	0	16	21	19	38	36	38

Table 5.10. NIBAS: comparison of marginal distributions concerning survey data Y

(1) Comparison of marginal distributions

As described in detail in Section 5.4.2 we calculate the test on the equality of proportions of zeros, perform the Kolmogorov–Smirnov goodness-of-fit test (KS) as well as the Wilcoxon signed rank test, and a standard two-sample paired t-test after having removed units with observed and imputed ties. Since we produce $m = 5$ imputed data sets, we calculate each statistic for every data set comparing the imputed distributions with the original distributions and count the number of times out of five that the null hypothesis is not rejected. For the imputation of X in file A the results are listed in Table 5.10. Notice that testing the proportions of zeros being the same is only valid when zeros are generated. Since the variables X_{20} and X_{21} are generated without zeros by NIBAS, no test results are given in Table 5.10 for those variables.

Obviously the distributional differences reported by the Kolmogorov–Smirnov test are affected by the proportions of zeros. Removing them yields a better fit shown by the Wilcoxon signed rank test and the paired t-test for the remaining continuous variables throughout every imputation in most of the cases. Furthermore the auxiliary file is of little help here.

Notice that the different tests sometimes lead to contradictory inference; for example, see variable Y_{24} and compare the results of the Wilcoxon signed rank test and the paired t-test. Such discrepancies may be due to the fact that the continuous part of the Y variables is still skewed even after being transformed toward normality; see Section 5.2.1.

Sustaining the marginal distributions of the Y variables seems to be even more difficult. This already points at a "better" statistical match when variables X of file B are imputed in file A than imputing variables Y of file A in file B which may be due to a better normal approximation of the transformed X variables. Again little is gained by using the auxiliary file.

(2) Bivariate associations

From the imputed data set without any transformations the multiple imputation estimate of Spearman's r_S and of the z-transformed Pearson's r_P is calculated and tested for every combination of the X and Y variables in each statistically matched file A and B. The results of these tests are compared with the results derived from the originally complete files A and B as shown in Figure 5.10. The conclusions at the 5% significance level are summarized in Table 5.11.

The multiple imputation interval estimates are expected to produce wider intervals leading to conservative tests which obviously is supported by our results. Compared to the complete data, the significance tests based on the imputed files are not likely to produce false positives; i.e., they wrongly reject the null hypothesis. Either inference is the same or we will not reject the hypothesis of (unconditional) "independence", at least, of noncorrelation. Nevertheless, we correctly specify up to 73% (52%) of the inference concerning the association of the variables never jointly observed

	Based on Spearman's r_S					
	Impute X in A			Impute Y in B		
Inference	No a	a_{300}	a_{600}	No a	a_{300}	a_{600}
Correct	42	45	38	40	49	44
False pos.	0	0	0	1	1	1
False neg.	21	18	25	22	13	18
Contradiction	0	0	0	0	0	0

	Based on Pearson's r_P					
	Impute X in A			Impute Y in B		
Inference	No a	a_{300}	a_{600}	No a	a_{300}	a_{600}
Correct	43	43	42	31	41	36
False pos.	0	0	1	0	0	0
False neg.	20	20	20	32	22	27
Contradiction	0	0	0	0	0	0

Table 5.11. NIBAS: inference about bivariate associations of survey data X and Y

based on the complete file A (B), when no auxiliary file is used. The best improvement in getting more identical inference is achieved with the small auxiliary file; the larger file is less helpful.

(3) Conditional independence

Now the tests of bivariate conditional association of X_i and Y_j given the common variables $Z = z$ for $i = 15, 16, \ldots, 21$, $j = 22, 23, \ldots, 30$ are performed either based on the multiple imputation estimate of $\widehat{\beta}^{MI}_{YX.Z}$ or the usual estimate $\widehat{\beta}_{YX.Z}$ using the original data files. Notice that 37 combinations in file A and 47 combinations in file B are reported to be conditionally independent or, at least, conditionally uncorrelated in the original files. According to Figure 5.11 inference about the conditional associations is compared and the correct or incorrect specifications are counted. The results are given in Table 5.12.

Without auxiliary files we find agreement in nearly 59% (75%) of all cases in file A (B). Since we have set the conditional correlations to zero a priori, **all** of the conditional correlations are reproduced as zero. Again

Inference based	Impute X in A			Impute Y in B		
on $\widehat{\beta}_{YX.Z}$	No a	a_{300}	a_{600}	No a	a_{300}	a_{600}
Correct	37	48	43	47	37	43
False pos.	0	4	4	0	3	0
False neg.	26	11	16	16	23	20
Contradiction	0	0	0	0	0	0

Table 5.12. NIBAS: inference about conditional independence of survey data X and Y given $Z = z$

Statistic	Impute X in A No a	a_{300}	a_{600}	Impute Y in B No a	a_{300}	a_{600}
Mean dev.	66.77	334.58	224.70	62.36	232.11	136.12
Counts	5	0	0	5	0	0

Table 5.13. NIBAS: inference about the mean deviance of survey data X and Y given $Z = z$

we get false negatives rather than false positives; i.e., in nearly all of the cases the tests are conservative.

The calculation of the mean deviance $\frac{1}{5}\sum dev(Y||X|Z)^{(k)}$ and the number of times the null hypothesis is not rejected for the five imputed data sets are presented in Table 5.13. This procedure allows us to test whether all X and Y variables may be conditionally independent given the common variables $Z = z$ in the imputed files.

Using the prior of zero for the conditional association of X and Y given $Z = z$ reproduces zero quite well and, thus, the hypothesis of conditional independence of X and Y given $Z = z$ is not rejected for every imputed data set when no auxiliary file is used. If information from an auxiliary file is taken then conditional independence is no longer achieved. The conditional independence assumption can obviously be overcome by use of the auxiliary files; i.e., using auxiliary information tends to establish conditional association. Comparing both Tables 5.12 and 5.13 we see that we get more correct specified inference by using auxiliary information the smaller the number of conditionally independent variables X and Y given $Z = z$ is. Hence, the utility of the auxiliary file obviously depends on the quality of the estimates of the conditional association derived therefrom.

(4) Cross-tabulation fit

The 63 possible combinations of the X and Y variables are cross-tabulated whereas the distributions are categorized into $5 \cdot 5 = 25$ cells each. These $63 \cdot 25 = 1575$ cell frequencies from the imputed data are compared with the 1575 cell frequencies derived from the original data and the correlation coefficients are computed according to (5.18) for every imputation. For the worst and the best cross-tabulation fit out of 5 correlation coefficients the imputed frequencies are plotted against the original frequencies. The fit is better for the imputations in file A than those of file B.

Statistic	Impute X in A No a	a_{300}	a_{600}	Impute Y in B No a	a_{300}	a_{600}
$\frac{1}{m}\sum r^{(k)}$	0.9066	0.9060	0.9138	0.6737	0.6686	0.6581
$\frac{1}{m}\sum r_{KW}^{(k)}$	0.7092	0.7346	0.7135	0.7363	0.7136	0.7139

Table 5.14. NIBAS: cross-tabulation fit of survey data

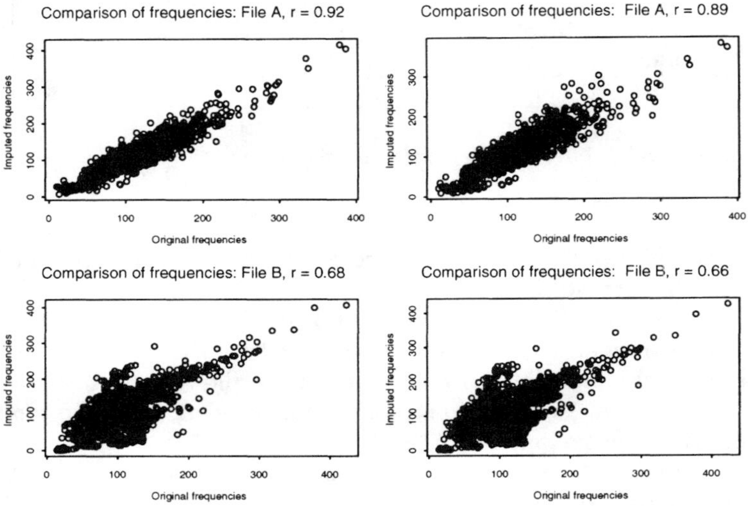

Figure 5.18. NIBAS: best and worst cross-tabulation fit

Figure 5.18 illustrates these plots when no auxiliary data are used. Using information from the auxiliary files does not improve the reproduction of the bivariate distributions much if it does at all; see Table 5.14.

Testing whether the bivariate categorized distributions could be drawn from the same underlying population only gives some good results for imputing the X variables in file A. Table 5.15 shows the minimum and the maximum number of times from the five imputed data sets the null hypothesis is rejected for each of the 63 X and Y combinations.

The test of homogeneity for every two categorized distributions of the X and Y variables in every imputed data set and the original data regards only a maximum of 42, i.e., $\frac{2}{3}$ of the bivariate distributions as similar. When imputing Y in file B we would reject the null hypothesis in nearly all cases. Obviously it is easier to preserve parameters of distributions than the distributions themselves although the reported cross-tabulation fit is rather high. Again we find only a small gain in using auxiliary information.

	Impute X in A			Impute Y in B		
Statistic	No a	a_{300}	a_{600}	No a	a_{300}	a_{600}
[min, max]	[30, 37]	[25, 33]	[28, 42]	[0, 0]	[0, 0]	[0, 2]

Table 5.15. NIBAS: inference about cross-tabulation fit of survey data

Inference based on regression	Impute X in A			Impute Y in B		
	No a	a_{300}	a_{600}	No a	a_{300}	a_{600}
Correct	181	177	178	166	163	167
False pos.	4	13	6	3	9	4
False neg.	13	8	14	29	26	27
Contradiction	0	0	0	0	0	0

Table 5.16. NIBAS: counting linear regression inferences based on survey data

(5) Linear regression inference

Now all possible multivariate linear regression models from each Y_j variable, $j = 22, 23, \ldots, 30$, on all (imputed) variables X and the common variables Z are calculated in file A as well as in file B where the Y values are the imputed ones. The results are compared with those achieved from the original complete files A and B. Counting the conclusions made at the 5% significance level leads to Table 5.16.

Here we find the best correspondence up to 91% in file A (81% in file B) of the inferences when no auxiliary file is used. Again we realize that the estimated values of the conditional correlation $\rho_{XY|Z}$ derived from the auxiliary files are less efficient than prior setting $\rho_{XY|Z}$ to zero. The auxiliary files increase the number of false positives which may be due to a decrease of quality of the estimated prior compared with the prior of zero. Due to the quite high amount of conditionally uncorrelated variables X and Y given $Z = z$, many regression coefficients are already regarded as zero which is well reproduced by the imputations under the prior of zero.

Using Generated Data

We take the random data set $(Z, X, Y) \sim N_{30}(\widehat{\mu}, \Sigma^\star)$ according to (5.13) and apply the same procedures as done before to these generated data.

File	KS			Wilcoxon			Paired t		
A	No a	a_{300}	a_{600}	No a	a_{300}	a_{600}	No a	a_{300}	a_{600}
X_{15}	5	4	5	5	3	5	5	2	5
X_{16}	5	5	5	5	4	5	5	4	5
X_{17}	5	5	5	5	4	1	5	3	1
X_{18}	5	5	4	4	3	2	4	3	1
X_{19}	4	5	4	2	3	3	4	3	2
X_{20}	4	5	5	5	5	5	5	5	5
X_{21}	3	3	4	2	2	5	2	2	5
$\sum$	31	32	32	28	24	26	30	22	24

Table 5.17. NIBAS: comparison of marginal distributions concerning generated data X

File	KS			Wilcoxon			Paired t		
B	No a	a_{300}	a_{600}	No a	a_{300}	a_{600}	No a	a_{300}	a_{600}
Y_{22}	3	4	5	4	2	5	4	2	5
Y_{23}	5	5	4	5	4	2	5	4	2
Y_{24}	5	5	5	5	5	5	5	5	5
Y_{25}	5	5	5	5	4	5	5	4	5
Y_{26}	1	4	4	3	3	4	3	3	4
Y_{27}	4	5	5	4	5	5	4	5	4
Y_{28}	5	5	5	5	4	5	5	5	5
Y_{29}	3	3	4	5	3	4	5	3	4
Y_{30}	5	4	4	5	5	2	5	4	2
$\sum$	32	40	41	41	35	37	41	35	36

Table 5.18. NIBAS: comparison of marginal distributions concerning generated data Y

(1) Comparison of marginal distributions

Our variables now come from a multivariate normal distribution; thus, testing the equality of proportions of zeros is meaningless here. Therefore we only perform the Kolmogorov–Smirnov (KS) goodness-of-fit test, the Wilcoxon signed rank test, and the standard two-sample paired t-test using the simulated data without further manipulations. Again we calculate each statistic for every data set comparing the imputed distributions with the original distributions and count the number of times out of five that the null hypothesis is not rejected. The results of the imputation of X in file A are shown in Table 5.17.

Imputing the Y variables in file B now yields much change as illustrated in Table 5.18. Depending on the test applied using the auxiliary file may improve or worsen sustaining the marginal distributions. In general the marginal distributions are restored quite well by each single imputation via NIBAS when the distributional assumptions are fulfilled.

(2) Bivariate associations

The conclusions at the 5% significance level concerning inference about the bivariate associations are summarized in Table 5.19. Under multiple imputation inference again the null hypothesis of independence between any X and Y is less often rejected if there is dependence than otherwise. The prior information of zero of the conditional correlation of X and Y given $Z = z$ now clearly misleads the inference, but the small auxiliary file yields up to 87% and more correct specifications. Hence, if there is no or only small conditional independence between the variables X and Y given $Z = z$ the use of auxiliary information may improve the imputation procedure remarkably.

	Based on Spearman's r_S					
	Impute X in A			Impute Y in B		
Inference	No a	a_{300}	a_{600}	No a	a_{300}	a_{600}
Correct	29	55	59	30	57	57
False pos.	0	1	1	1	3	1
False neg.	34	7	3	32	3	5
Contradiction	0	0	0	0	0	0

	Based on Pearson's r_P					
	Impute X in A			Impute Y in B		
Inference	No a	a_{300}	a_{600}	No a	a_{300}	a_{600}
Correct	29	55	59	29	54	54
False pos.	0	1	1	1	2	2
False neg.	34	7	3	33	7	7
Contradiction	0	0	0	0	0	0

Table 5.19. NIBAS: inference about bivariate associations of generated data X and Y

(3) Conditional independence

We consider the bivariate conditional independence of Y_j and X_i given the common variables $Z = z$ for $i = 15, 16, \ldots, 21$, $j = 22, 23, \ldots, 30$ using the imputed and the complete data for the two files. The results are displayed in Table 5.20. Clearly, the imputations under the prior assumption of conditional independence reproduce this prior assumption of zero. Thus, in the imputed data set the hypothesis of conditional independence is never rejected, leading to a high amount of false negatives. Here any auxiliary file will improve the inference drawn from the imputed data considerably, leading to over 80% of correct specifications.

Calculating the mean deviance $\frac{1}{5}\sum dev(Y||X|Z)^{(k)}$ and counting the number of not rejecting the null hypothesis for the five imputed data gives the same results as shown in Table 5.21.

The prior assumption of conditional independence is reproduced in the imputed data whereas any auxiliary file leads to rejecting the hypothesis of X and Y being independent given $Z = z$.

Inference based	Impute X in A			Impute Y in B		
on $\widehat{\beta}_{YX.Z}$	No a	a_{300}	a_{600}	No a	a_{300}	a_{600}
Correct	9	50	51	9	52	51
False pos.	0	3	3	0	3	3
False neg.	54	10	9	54	6	8
Contradiction	0	0	0	0	2	1

Table 5.20. NIBAS: inference about conditional independence of generated data X and Y given $Z = z$

Statistic	Impute X in A			Impute Y in B		
	No a	a_{300}	a_{600}	No a	a_{300}	a_{600}
Mean dev.	66.72	8080.07	11188.93	59.34	8157.58	11122.73
Counts	4	0	0	5	0	0

Table 5.21. NIBAS: inference about the mean deviance of generated data X and Y given $Z = z$

(4) Cross-tabulation fit

Again we find that the prior of zero is highly misleading; see Table 5.22. The correlation coefficients are high throughout when an auxiliary file is used. The direction in which the matching is performed makes no difference now. The results are more or less the same whether X is imputed in file A or Y is imputed in file B.

Compared with the results based on the survey data it should be noticed that even the bivariate distributions seem to be reproduced well when use of the auxiliary file is made. In many cases the χ^2 test will never reject the hypothesis that the bivariate categorized distributions of X and Y in the original and the imputed data set come from the same population; see Table 5.23.

(5) Linear regression inference

The ordinary linear model regressing each Y variable on all X and the common variables Z is estimated based on the generated data. The conclusions made at the 5% significance level are compared; the results of the inferences are shown in Table 5.24.

Inference concerning the regression coefficients is misleading when based on the prior of zero. In this case too often the regression coefficients are reported to be zero due to the conditional independence assumed a priori. By use of an auxiliary file we achieve 84% (file A) and 72% (file B) of identical conclusions. Here the tests are less conservative compared to other

Statistic	Impute X in A			Impute Y in B		
	No a	a_{300}	a_{600}	No a	a_{300}	a_{600}
$\frac{1}{m}\sum r^{(k)}$	0.3615	0.8252	0.8175	0.3599	0.8115	0.8046
$\frac{1}{m}\sum r_{KW}^{(k)}$	0.4122	0.8474	0.8459	0.4143	0.8438	0.8334

Table 5.22. NIBAS: cross-tabulation fit of generated data

Statistic	Impute X in A			Impute Y in B		
	No a	a_{300}	a_{600}	No a	a_{300}	a_{600}
[min, max]	[31, 35]	[59, 63]	[60, 63]	[31, 36]	[57, 62]	[56, 63]

Table 5.23. NIBAS: inference about cross-tabulation fit of generated data

Inference based on regression	Impute X in A			Impute Y in B		
	No a	a_{300}	a_{600}	No a	a_{300}	a_{600}
Correct	92	157	167	93	140	143
False pos.	12	12	4	1	10	6
False neg.	88	27	26	100	47	49
Contradiction	6	2	1	4	1	0

Table 5.24. NIBAS: counting linear regression inferences based on generated data

situations discussed before. In this case we observe the highest amount of false positives.

Summary

Within the survey data a high amount of variables X and Y being conditionally independent given the common variables $Z = z$ is already incorporated. Thus, the prior information assuming conditional independence, i.e., $R_{XY|Z}$ is set to zero, is particularly suitable in this context. Obviously the noniterative multivariate model allows a quite efficient statistical matching when the common variables are well selected. Estimating this prior from an auxiliary file seems to be less efficient unless the auxiliary file is sufficiently large or of "good" quality as shown in comparison with the generated data. The better the model assumption of normality for the variables to be imputed is met the better the auxiliary information can be exploited. In our cases we found less or no improvement with the auxiliary file of size $n = 300$ when the survey data are used.

Moreover, we get better results with the survey data when the X variables are imputed in file A than imputing the Y variables in file B. This might be due to the fact that the normal approximation is achieved better for the X variables. Using the simulated data we find no difference concerning the directions of the match.

Comparing the cross-tabulation fit r_{KW} with the results presented by Kamakura and Wedel (1997), we calculate rather similar values. The correlation coefficients are regarded as rather high although it is obviously difficult to preserve the joint distributions of the variables not jointly observed. Again the skewed distributions and the large amount of zeros are a hindrance to achieving the second level of validity. Assuming normality and using auxiliary information the categorized joint distributions are well preserved.

Investigating the imputations produced by NIBAS we find only few differences between them although the fraction of missing information is about 50%. This might be due to the model's ability to produce rather stable imputations. Using the multiple imputation estimates to compare the inferences based on the imputed data with the inference drawn from the actual data only leads to a higher amount of false negatives and a slight reduction of the false positives. This is affected by the multiple imputation interval estimates, which produce wider confidence intervals than the usual complete case estimates to account for the imputation's uncertainty. For the sake of simplicity one might stay with one single imputation from the Bayesian model at least with this sort of data making tests less conservative here.

Another advantage of this model is its computational convenience. The imputations are calculated within seconds even for the large survey data set.

5.5.2 Data Augmentation

Again $m = 5$ imputations are created using the normal model and the transformations as described in Section 5.2.1; we are imputing the model with 46 variables.[9] When the auxiliary file is used, its data sets are simply added to the concatenated data files (see figure 5.19) and the noninformative prior distribution of the parameters is assumed according to the description in Section 4.6.2.

Remember that the data are divided into a binomial and a continuous part; to the latter is applied the log as well as the logit transformation to the positive values of each variable X and Y to achieve normality. Figure 5.4 illustrates this transformation and shows that the relationship among the encoded variables W and V is never identified even if auxiliary information is available. By means of the auxiliary files the association of X and Y becomes estimable but the model is still unidentified.

When data augmentation is performed, first initial parameter estimates are obtained by running the EM algorithm implemented in NORM 2.02 until convergence is achieved after 670 iterations. Since the parameters of the conditional association of X and Y given $Z = z$ are unidentified, the EM converges quite quickly compared to the large amount of missing data, but the resulting estimates are not unique. This can be shown when another run is performed using different starting values; for an example, see Tables 4.2 and 4.3. Since we do not intend to draw any inference from the EM solutions directly, we take the EM estimates as starting values for the data augmentation algorithm as described in Section 4.6. Running data augmentation itself with an unidentified model could drift some parameters to extremely large values or produce numeric overflow. A way to handle this problem is to introduce a small amount of information using, e.g., an informative or a ridge prior. For a detailed description of the informative prior or the ridge prior see Section 4.6.2. Using the ridge prior provided by

[9]The S-PLUS library NORM provided by Schafer (1997) is limited to a maximum of 30 variables, but the standalone MS-Windows^{TM} version NORM 2.02 is able to handle this data set; see Schafer (1999b).

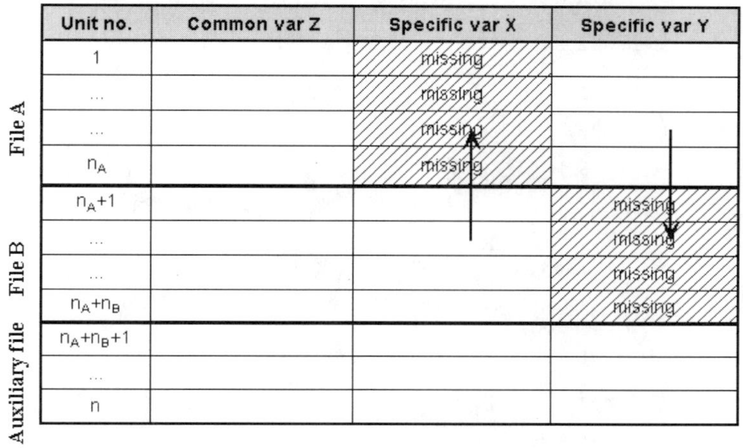

Figure 5.19. NORM: using auxiliary information

NORM 2.02, with $\tau = 0$ and $\Lambda^{-1} = \nu I$ for some hyperparameter ν set to a small positive value, should stabilize computations.[10] To check the convergence of the Markov chain we run 3000 cycles of the data augmentation algorithm under a ridge prior with the hyperparameter ν set to 2 which Schafer and Olsen (1999) report to be a good choice. This prior smooths the estimated correlations a bit towards a model of mutual independence among all variables; see Schafer (1997), p. 207.

Imputed values of the binary variables are rounded. Judging the time series plot provided by NORM 2.02 for the "worst linear function" of the parameters suggests that a burn-in period of 2000 cycles of the data augmentation algorithm may be enough to achieve independence of the chain from its starting values. Figure 5.20 displays this function and the sample autocorrelation function estimated therefrom. Convergence with respect to the worst linear function strengthens the evidence for global convergence of the chains; for details see Schafer (1997), pp. 129–131.

The lag-k autocorrelation of this function estimated from 2000 cycles of data augmentation obviously dies down after a lag of about 150; thus imputations generated at every $k = 200$ cycles should be approximately independent. Hence we decide to run a burn-in period of 2000 cycles and then another 1000 cycles stopping every 200 cycles to impute the missing values; i.e., $m = 5$ imputations are created from the 2200th, 2400th, . . . , 3000th run. When the auxiliary file a_{300} is added to the data this works quite well, too. With the larger auxiliary file a_{600} we recognize that the autocorrelation of the worst linear function dies down more slowly; thus

[10] An extensive explanation of the usage of NORM 2.02 under an unidentified model and a suitable parameter setting is given by Schafer and Olsen (1999).

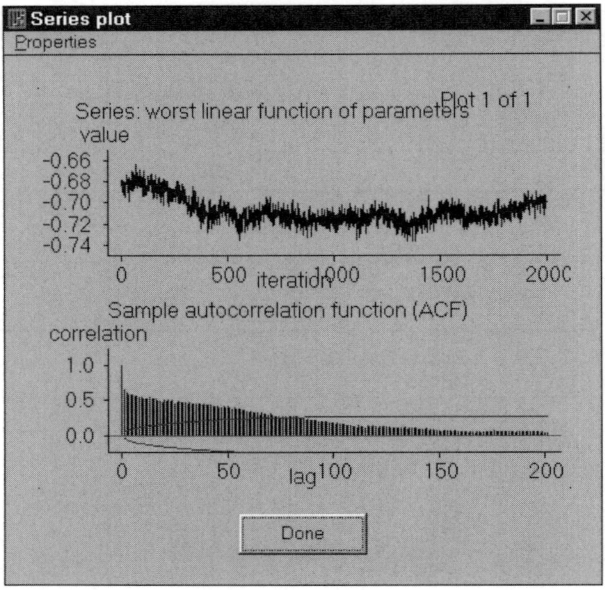

Figure 5.20. NORM: worst convergence of a Markov chain

we stop every 500 cycles from another 2500 cycles for creating the $m = 5$ imputations.

The imputed distributions of the first three X variables are presented in Figure 5.21. Then they are compared with the original data when no auxiliary file is used.

The structure of the imputed data of X with $n_A = 2748$ in file A is obviously quite similar to the structure of the original data; see Table 5.25. Again no zero values are generated for the variables X_{20} and X_{21}.

Notice that unrealistic large values for Y are redrawn as described in Section 5.2.1. Sometimes we do not achieve smaller values of Y within its

Variable	% Zeros		Mean		Standard error	
X_i	p_i^{org}	$\frac{1}{5}\sum p_i^{(k)imp}$	$\overline{X}_i^{org}$	$\frac{1}{5}\sum \overline{X}_i^{(k)imp}$	S_i^{org}	$\frac{1}{5}\sum S_i^{(k)imp}$
X_{15}	0.31	0.35	0.09	0.09	0.14	0.15
X_{16}	0.20	0.23	0.17	0.16	0.21	0.21
X_{17}	0.12	0.12	0.28	0.27	0.26	0.27
X_{18}	0.14	0.16	0.18	0.18	0.19	0.21
X_{19}	0.05	0.03	0.42	0.43	0.28	0.29
X_{20}	0.02	0.00	0.43	0.41	0.25	0.26
X_{21}	0.01	0.00	0.49	0.48	0.25	0.26

Table 5.25. NORM: marginal distributions of imputed variables X_{15}, X_{16}, and X_{17}

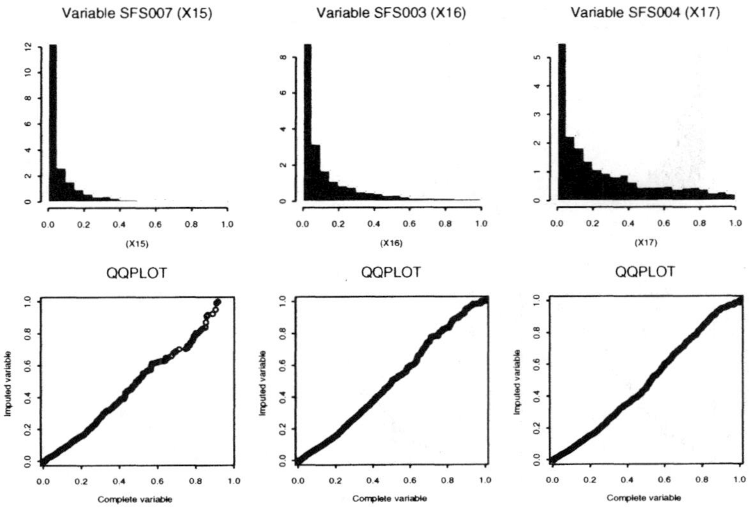

Figure 5.21. NORM: histogram and qqplot of imputed variables X_{15}, X_{16}, and X_{17}

bounds after the third time of redrawing; then the maximum feasible value of Y is taken as the imputation. Thus, the qqplots in Figure 5.22 display more differences between the original and the imputed data due to these large values.

The comparison of the imputed and original variables Y in file B when no auxiliary file is used displays greater differences concerning the estimated means and standard errors; see Table 5.26.

Variable	% Zeros		Mean		Standard error	
Y_i	p_i^{org}	$\frac{1}{5}\sum p_i^{(k)imp}$	$\overline{Y}_i^{org}$	$\frac{1}{5}\sum \overline{Y}_i^{(k)imp}$	S_i^{org}	$\frac{1}{5}\sum S_i^{(k)imp}$
Y_{22}	0.07	0.05	10.72	12.16	10.83	18.02
Y_{23}	0.22	0.24	33.66	39.25	77.78	121.89
Y_{24}	0.27	0.27	6.22	6.32	9.88	11.96
Y_{25}	0.05	0.03	8.22	10.24	8.56	29.87
Y_{26}	0.37	0.39	1.38	1.26	2.48	2.21
Y_{27}	0.09	0.07	8.95	10.27	8.14	17.47
Y_{28}	0.38	0.37	1.80	1.77	3.40	4.04
Y_{29}	0.32	0.31	0.04	0.02	0.27	0.04
Y_{30}	0.32	0.31	1.55	1.33	3.83	3.45

Table 5.26. NORM: marginal distributions of imputed variables Y_{22}, Y_{23}, and Y_{24}

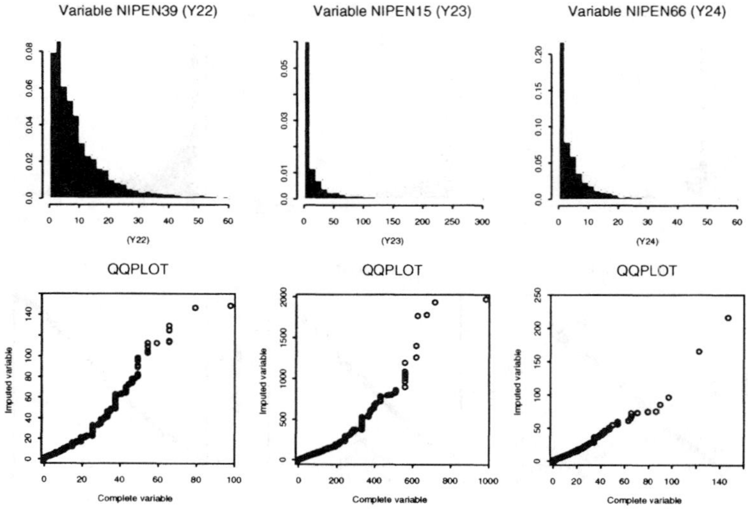

Figure 5.22. NORM: histogram and qqplot of imputed variables Y_{22}, Y_{23}, and Y_{24}

Goodness-of-Fit

To get starting values of the parameters in the data augmentation I-step the EM algorithm is run first. Convergence of the EM algorithm is achieved after 670 iterations when no auxiliary information is taken, and after 668 (505) iterations when auxiliary files of size $n = 300$ or $n = 600$ are used, respectively. The default starting values of the EM algorithm use the observed data means and standard deviations; the starting values for all covariances are set to zero. Then we run 2000 cycles of the data augmentation as the burn-in period and another 1000, respectively, 2500, cycles to compute $m = 5$ independent imputations using the ridge prior with the hyperparameter ν set to 2.

(1) Comparison of the marginal distributions

Again the two-sample tests are performed concerning the differences of the proportions of zeros as well as the means and the variances in the original and the imputed data set of files A and B, respectively. The results are shown in Table 5.27. No zero values are imputed for variables X_{20} and X_{21}; thus, no test results are given for the proportions of zeros. Similar to the imputations produced by NIBAS, the proportions of zeros are often quite different in the imputed data set and the original data set. When the zero values are removed and only the remaining continuous parts of the variables are compared then the marginal distributions are preserved, at least, in the majority of the imputed data sets.

File	Equal % zeros			KS			Wilcoxon			Paired t		
A	No a	a_{300}	a_{600}	No a	a_{300}	a_{600}	No a	a_{300}	a_{600}	No a	a_{300}	a_{600}
X_{15}	0	0	1	0	0	0	4	5	5	4	3	5
X_{16}	0	0	2	0	1	0	5	4	4	4	4	4
X_{17}	5	5	5	0	3	0	2	4	1	2	4	1
X_{18}	2	3	3	0	0	0	5	5	3	5	4	1
X_{19}	2	0	0	3	0	0	3	4	1	2	3	3
X_{20}	–	–	–	0	0	2	0	1	3	1	3	3
X_{21}	–	–	–	0	0	1	1	5	3	4	5	3
$\sum$	9	8	11	3	4	3	20	28	20	22	26	20

File	Equal % zeros			KS			Wilcoxon			Paired t		
B	No a	a_{300}	a_{600}	No a	a_{300}	a_{600}	No a	a_{300}	a_{600}	No a	a_{300}	a_{600}
Y_{22}	0	2	0	0	0	0	4	3	0	4	5	1
Y_{23}	3	2	1	0	0	0	5	5	3	5	5	4
Y_{24}	5	0	2	0	0	0	3	2	3	2	2	4
Y_{25}	0	0	0	0	0	0	1	1	3	3	2	2
Y_{26}	5	4	1	0	0	0	4	4	5	5	2	5
Y_{27}	1	5	0	0	0	0	0	3	2	2	4	3
Y_{28}	5	5	3	0	0	0	5	5	4	2	4	4
Y_{29}	4	2	0	0	0	0	3	1	1	1	1	3
Y_{30}	5	1	0	0	0	0	3	5	5	1	5	4
$\sum$	28	21	7	0	0	0	28	29	26	25	30	30

Table 5.27. NORM: comparison of marginal distributions concerning survey data Y

	Based on Spearman's r_S					
	Impute X in A			Impute Y in B		
Inference	No a	a_{300}	a_{600}	No a	a_{300}	a_{600}
Correct	35	28	27	34	31	33
False pos.	1	0	1	1	0	0
False neg.	27	35	35	25	32	30
Contradiction	0	0	0	3	0	0

	Based on Pearson's r_P					
	Impute X in A			Impute Y in B		
Inference	No a	a_{300}	a_{600}	No a	a_{300}	a_{600}
Correct	28	28	37	27	28	20
False pos.	4	0	0	2	0	0
False neg.	31	35	26	33	35	43
Contradiction	0	0	0	1	0	0

Table 5.28. NORM: inference about bivariate associations of survey data X and Y

Again we see that the auxiliary information does not always help to sustain the marginal distributions. The results reported from the Wilcoxon signed rank test are better here than with NIBAS although the results given by the paired t-test are the opposite. Obviously both imputation procedures are scarcely influenced by auxiliary information to achieve a better reproduction of the marginal distributions.

(2) Bivariate associations
The correlation coefficients according to Spearman and Pearson are calculated and tested for each of the 63 combinations of X and Y. We summarize the results of the tests at the 5% significance levels for the imputation of X in file A and Y in file B in Table 5.28. The tests based on the augmented data are conservative, similar to those based on the noniterative multivariate imputations although there are fewer correct specified inferences here. We find only up to 59% in file A (46% in file B) of the inference is correctly specified when no auxiliary file is used. In most of the cases the auxiliary information is of less help. The ridge prior as well as the auxiliary files tend to establish (unconditional) independence or, at least, no correlation between X and Y. Thus, the number of false negatives is rather high here.

(3) Conditional independence
We consider the test of bivariate conditional association of X_i and Y_j given the common variables Z for $i = 15, 16, \ldots, 21$, $j = 22, 23, \ldots, 30$ using the multiple imputation estimate from the imputed and the usual estimate from the complete data for the two matching directions. The results are given in Table 5.29.

Inference based on $\widehat{\beta}_{YX.Z}$	Impute X in A No a	a_{300}	a_{600}	Impute Y in B No a	a_{300}	a_{600}
Correct	33	51	41	42	37	41
False pos.	8	1	1	7	0	0
False neg.	20	11	21	13	26	22
Contradiction	2	0	0	1	0	0

Table 5.29. NORM: inference about conditional independence of survey data X and Y given $Z = z$

Contrary to the noniterative imputation procedure NIBAS, where we imputed conditional independence due to the chosen prior, we now observe less conditional independence with the ridge prior and even less when auxiliary information is added to the data. As was to be expected the ridge prior smooths the unconditional associations of X and Y towards zero; hence, the conditional associations now reflect the fusion covariances that were produced by the traditional approaches because then $\Sigma_{XY|Z} \approx \Sigma_{XZ}\Sigma_{ZZ}^{-1}\Sigma_{ZY} = \widetilde{cov}(X,Y)$ holds. Nevertheless, inferences agree up to 56% in file A (67% in file B). The mean deviance $\frac{1}{5}\sum dev(Y||X|Z)^{(k)}$ and the number of times the null hypothesis is not rejected for the five imputed data sets are presented in Table 5.30.

Again the ridge prior leads to unconditional noncorrelation rather than conditional noncorrelation. Under the ridge prior the hypothesis of X and Y being conditionally independent given $Z = z$ is clearly rejected for every imputed data set. Using auxiliary information and, thus, the usual noninformative prior instead of the ridge prior seems to improve the imputation process.

(4) Cross-tabulation fit

To discuss the preservation of the joint distributions in the imputed data sets the categorized distributions are cross-tabulated and the overall fit is computed by means of the correlation coefficients according to (5.18). With NORM we find the fit between the cell frequencies from the matched data and those from the complete data less perfect as shown in Figure 5.23 and Table 5.31.

Only a few differences between the correlation coefficient computed for the cell frequencies and r_{KW} are displayed. No matching direction obviously

Statistic	Impute X in A No a	a_{300}	a_{600}	Impute Y in B No a	a_{300}	a_{600}
Mean dev.	2422.08	547.48	324.85	1579.96	351.67	194.63
Counts	0	0	0	0	0	0

Table 5.30. NORM: inference about the mean deviance of survey data X and Y given $Z = z$

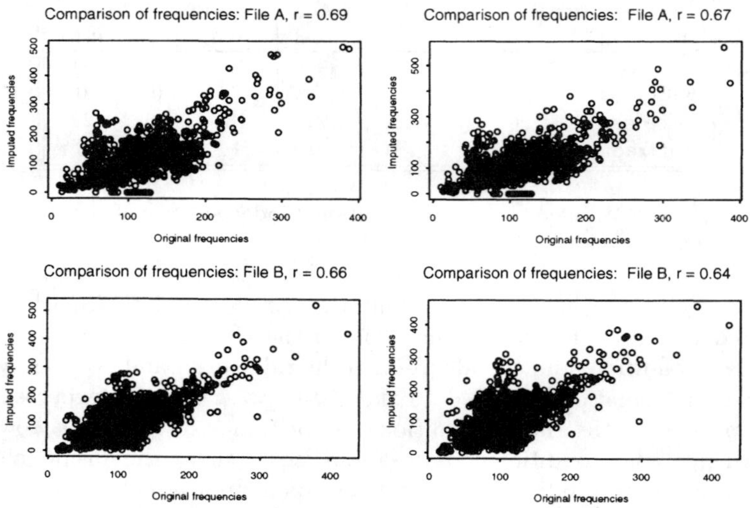

Figure 5.23. NORM: best and worst cross-tabulation fit

is better than the other: the imputation of Y in file B is handled as well as the imputation of X in file A. Table 5.32 shows the number of times the tests of homogeneity of the bivariate categorized distributions are not rejected. The results concerning the imputation of Y in file B are similar to those produced by NIBAS but NIBAS is more efficiently imputing X in file A. However, the preservation of the joint distributions is difficult to achieve by each imputation technique.

	Impute X in A			Impute Y in B		
Statistic	No a	a_{300}	a_{600}	No a	a_{300}	a_{600}
$\frac{1}{m}\sum r^{(k)}$	0.6796	0.6254	0.6956	0.6508	0.7302	0.7441
$\frac{1}{m}\sum r_{KW}^{(k)}$	0.6299	0.6548	0.6828	0.6152	0.6896	0.7085

Table 5.31. NORM: cross-tabulation fit of survey data

	Impute X in A			Impute Y in B		
Statistic	No a	a_{300}	a_{600}	No a	a_{300}	a_{600}
[min, max]	[0, 2]	[0, 0]	[4, 9]	[0, 0]	[0, 0]	[0, 0]

Table 5.32. NORM: inference about cross-tabulation fit of survey data

Inference based on regression	Impute X in A			Impute Y in B		
	No a	a_{300}	a_{600}	No a	a_{300}	a_{600}
Correct	158	186	180	155	162	161
False pos.	14	4	2	7	2	2
False neg.	26	8	16	36	34	35
Contradiction	0	0	0	0	0	0

Table 5.33. NORM: counting linear regression inferences based on survey data

(5) Linear regression inference

The results concerning the linear regression of the Y variables on all X variables and the common variables Z are given in Table 5.33. There the conclusions made at the 5% significance level are counted. When analyzing regression inference we get more coinciding inferences throughout when auxiliary information is used. For file A even 91 to 95% of the inferences are identical then, in file B about 82%. Due to the ridge prior more false positives are produced; i.e., sometimes conditional association is reported after imputation although there is none in the original data set.

Using Generated Data

We use the random draw for $(Z, X, Y) \sim N_{30}(\widehat{\mu}, \Sigma^{\star})$ already taken and perform the same investigations as done before with this generated data set. The EM algorithm converges very quickly now after 26 iterations without additional information and with 333 or 155 iterations when the auxiliary files of size 300 or 600 are used. Again we run 2000 cycles as a burn-in period with even the worst linear function dying down soon. From another 1000 cycles we stop every 200 cycles to get the $m = 5$ imputations. The hyperparameter ν is also set to 2 when no auxiliary information can be used; otherwise the usual noninformative prior is assumed. Notice that with additional information the model parameters are completely estimable now.

File	KS			Wilcoxon			Paired t		
A	No a	a_{300}	a_{600}	No a	a_{300}	a_{600}	No a	a_{300}	a_{600}
X_{15}	4	4	4	4	4	4	4	4	4
X_{16}	5	4	5	5	4	5	5	4	5
X_{17}	4	3	4	2	0	3	2	0	3
X_{18}	4	5	5	2	2	4	2	2	4
X_{19}	5	5	5	5	5	5	4	5	5
X_{20}	5	3	3	5	1	3	5	1	3
X_{21}	1	4	4	1	2	4	1	2	4
$\sum$	28	28	30	24	18	28	23	18	28

Table 5.34. NORM: comparison of marginal distributions concerning generated data X

File B	KS No a	KS a_{300}	KS a_{600}	Wilcoxon No a	Wilcoxon a_{300}	Wilcoxon a_{600}	Paired t No a	Paired t a_{300}	Paired t a_{600}
Y_{22}	3	5	5	3	5	4	3	5	4
Y_{23}	5	5	5	5	5	4	5	5	3
Y_{24}	4	5	5	3	5	5	3	5	5
Y_{25}	5	4	5	4	4	4	4	4	4
Y_{26}	3	5	5	2	5	5	2	4	5
Y_{27}	4	3	5	5	3	5	5	3	5
Y_{28}	5	5	5	5	4	4	4	4	5
Y_{29}	4	4	5	4	5	5	4	4	5
Y_{30}	4	3	5	4	2	4	4	2	4
$\sum$	37	39	45	35	38	40	34	36	40

Table 5.35. NORM: comparison of marginal distributions concerning generated data Y

(1) Comparison of marginal distributions

Table 5.34 lists the results of the comparison of the marginal distributions when variables X are imputed in file A. The results based on imputing the Y variables in file B are given in table 5.35.

Since there are no proportions of zeros now and the assumption of normality is fulfilled, the Kolmogorov–Smirnov test only rejects the null hypothesis in less than 20% of the cases. Contrary to NIBAS, especially the large auxiliary file is quite useful for the imputation procedure NORM to preserve the marginal distributions more often in each single imputation.

	Based on Spearman's r_S					
	Impute X in A			Impute Y in B		
Inference	No a	a_{300}	a_{600}	No a	a_{300}	a_{600}
Correct	26	50	54	24	53	52
False pos.	1	0	0	3	0	0
False neg.	34	13	9	34	10	11
Contradiction	2	0	0	2	0	0

	Based on Pearson's r_P					
	Impute X in A			Impute Y in B		
Inference	No a	a_{300}	a_{600}	No a	a_{300}	a_{600}
Correct	24	48	53	24	52	53
False pos.	1	0	0	2	0	0
False neg.	36	15	10	35	11	10
Contradiction	2	0	0	2	0	0

Table 5.36. NORM: inference about bivariate associations of generated data X and Y

Inference based on $\widehat{\beta}_{YX.Z}$	Impute X in A No a	a_{300}	a_{600}	Impute Y in B No a	a_{300}	a_{600}
Correct	23	45	49	21	47	46
False pos.	2	0	0	3	0	1
False neg.	36	18	14	37	16	16
Contradiction	2	0	0	2	0	0

Table 5.37. NORM: inference about conditional independence of generated data X and Y given $Z = z$

(2) Bivariate associations

The conclusions concerning the unconditional bivariate association of X and Y at the 5% significance level are summarized in Table 5.36. The ridge prior tends to establish independence between X and Y in the imputed data set; in two cases we even get contradictory inference. Auxiliary information improves the imputation procedure considerably: about 84% of the possible associations are correctly specified after the imputation.

(3) Conditional independence

Comparing the inference concerning conditional independence of Y_j and X_i given the common variables Z for $i = 15, 16, \ldots, 21$, $j = 22, 23, \ldots, 30$ based on the imputed and the complete data for the two files yields the results listed in Table 5.37.

Under the ridge prior only about 35% are correctly specified; the others are regarded as conditionally independent although there is association reported. By means of the auxiliary file the correct specifications rise to 80% in file A. With these generated data the ridge prior works better than the prior of zero with the noniterative procedure NIBAS. On the other hand NIBAS makes better use of the auxiliary files. In general the results are quite similar to those achieved from the survey data. As in Table 5.30, in Table 5.38 it is shown that NORM always leads to a rejection of the hypothesis of conditional independence of the variables X and Y given $Z = z$.

(4) Cross-tabulation fit

Similar to the results produced by NIBAS under the prior of zero imputations by NORM based on the ridge prior do not preserve the joint distributions well. Table 5.39 lists the correlation coefficients calculated

Statistic	Impute X in A No a	a_{300}	a_{600}	Impute Y in B No a	a_{300}	a_{600}
Mean dev.	21814	10372	10368	21947	10331	10563
Counts	0	0	0	0	0	0

Table 5.38. NORM: inference about the mean deviance of generated data X and Y given $Z = z$

Statistic	Impute X in A No a	a_{300}	a_{600}	Impute Y in B No a	a_{300}	a_{600}
$\frac{1}{m}\sum r^{(k)}$	0.5356	0.8177	0.8271	0.5367	0.8207	0.8175
$\frac{1}{m}\sum r_{KW}^{(k)}$	0.5476	0.8480	0.8482	0.5507	0.8505	0.8434

Table 5.39. NORM: cross-tabulation fit of generated data

Statistic	Impute X in A No a	a_{300}	a_{600}	Impute Y in B No a	a_{300}	a_{600}
[min, max]	[26, 30]	[60, 62]	[63, 63]	[21, 30]	[59, 63]	[60, 63]

Table 5.40. NORM: inference about cross-tabulation fit of generated data

from the categorized bivariate distributions of the imputed and the original data set.

The cross-tabulation fit produced by NORM behaves as does the fit produced by the noniterative multivariate imputation procedure NIBAS. If auxiliary information is available then the tests of homogeneity of the categorized distributions of X and Y in the imputed and the original sample are seldom rejected; see Table 5.40. Obviously these tests are heavily influenced by the skewed distributions of the survey data. Under more ideal conditions the imputation procedures seem to achieve the second level of validity quite well.

(5) Linear regression inference

Finally, the conclusions made at the 5% significance level of the linear regression are compared and counted; see Table 5.41.

The ridge prior can be quite misleading when regression inference is done. This is a plausible result because unconditional rather than conditional independence is introduced by the ridge prior. In some cases even a contradictory decision is reported from the imputed data set compared with the original data set. However, the tests based on the imputed data are rather conservative, tending not to reject the null hypothesis of no influence of a regression coefficient although we may find a significant influence

Inference based on regression	Impute X in A No a	a_{300}	a_{600}	Impute Y in B No a	a_{300}	a_{600}
Correct	110	147	158	91	125	150
False pos.	5	4	5	12	4	1
False neg.	78	47	35	91	69	47
Contradiction	5	0	0	4	0	0

Table 5.41. NORM: counting linear regression inferences based on generated data

in the original data. Making use of auxiliary information always improves the inference based on the multiple imputation estimates.

Summary

The results are not as good as those obtained when the noniterative multivariate procedure NIBAS is used. Using a ridge prior is necessary to ensure the convergence of the chain, but tends to mislead the inference based on the imputed data. Auxiliary information can be used very easily just by adding the information to the data. Even a small amount of completely observed data may partly improve the estimation results.

Note that the ridge prior smooths the unconditional associations toward zero, whereas the prior of the non-iterative multivariate model sets the conditional associations to zero. Clearly we retrieve this in our results. When the data augmentation algorithm is applied to the survey data no matching direction is obviously better than the other.

The imputations themselves are obviously less stable than those produced by the noniterative multivariate procedure, producing wider confidence intervals and, thus, more false negatives. Therefore tests are even more conservative.

Computation of the data augmentation imputations is easy, since NORM 2.02 as well as the latest version NORM 2.03, is simple to handle. With the survey data set, the burn-in period took more than 30 minutes to be computed. Redrawing unrealistic large values is a bit uncomfortable since one has to switch between a programming language and NORM 2.02. Nevertheless, every setting used or calculated by NORM 2.02 can be stored, facilitating further computations. The preparation of the data by finding suitable transformations is also supported by NORM 2.02.

5.5.3 Iterative Imputations by Chained Equations

Finally we apply the most flexible imputation technique available called MICE, which stands for multiple imputations by chained equations.[11] With MICE iterative univariate imputations are performed reducing the problem of multivariate missing data to a simpler regression modeling making use of a Gibbs sampling procedure. Despite the attractive features of this method, its statistical properties are not well understood because the joint distribution of the variables to be imputed may not actually exist and convergence of the Gibbs sampler may not be achieved in those cases. In our setting we treat the X and Y variables as more or less univariate normally distributed imputing the missing values using linear regression analysis, i.e., by means of the predictive mean matching procedure. Since we do not

[11]The S-PLUS library MICE has been available on the Internet since June 2000; see Web site www.multiple-imputation.com.

	Unit no.	Common var Z	Y	Ypred
File A	1	z1	1	1.5
	2	z2	5	3.4
	3	z3	0	0.1
	4	z4	0	0.3
	5	z5	12	7
File B	6	z6	NA	3.2
	7	z7	NA	1.6
	8	z8	NA	8
	9	z9	NA	0.1

Figure 5.24. MICE: a simple illustration of predictive mean matching

use the transformed data with their binomial parts, the joint distribution of the variables being imputed may exist at least theoretically. Van Buuren and Oudshoorn (2000) report satisfactory performance and good results with as few as 5 or 10 iterations; thus we decide to stay with 10 iterations. Notice that similar to our noniterative multivariate model no iterations are needed to achieve independence in the imputations themselves as the data augmentation algorithms do. The Gibbs sampler is used only to create independence between the univariate posterior distributions of the regression coefficients since these parameters are not drawn at once from their multivariate posterior distribution as is done in the noniterative multivariate model.[12] Using the standard procedure as implemented in MICE without transforming our data we create $m = 5$ multiple imputations again. Although the matching procedure itself is not based on a random process but on the predictive mean matching, different imputations are created due to the underlying Bayesian regression model. Notice that within the predictive matching the values that are imputed by MICE are observed values and not produced by random draws. Although the underlying model is a Bayesian-based model as described in Section 4.7, the imputed values are finally derived by a nearest neighbor search based on the predicted values generated in accordance with the model. Figure 5.24 illustrates the procedure used by predictive mean matching on a simplified example.

When the auxiliary file is used again it is simply added to the data as previously done with the data augmentation procedure. The univariate distributions of the imputed variables X in file A are shown in Figure 5.25. The structure of the imputed data of X with $n_A = 2748$ in file A is listed in Table 5.42. Few differences can be seen between the moments estimated from the original and the imputed data sets.

[12]Extensive descriptions of the MICE library and its functionality can be found by the interested reader in the TNO reports by van Buuren and Oudshoorn (1999) and Oudshoorn et al. (1999), and in the MICE V1.0 User's Manual.

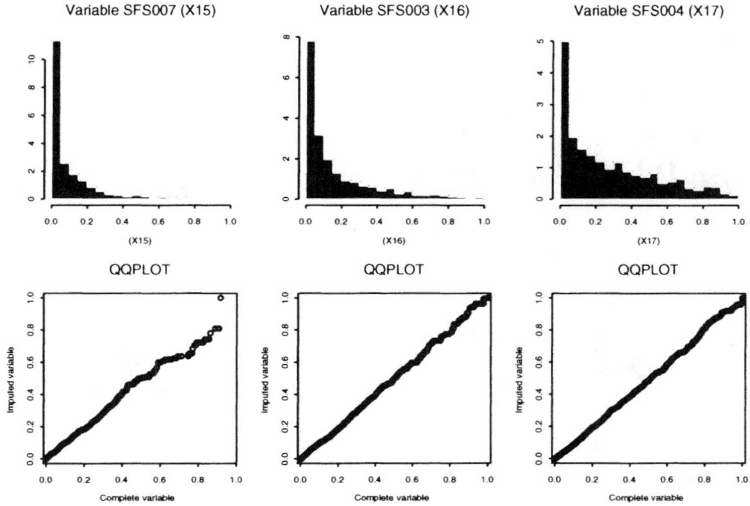

Figure 5.25. MICE: histogram and qqplot of imputed variables X_{15}, X_{16}, and X_{17}

The impact of the matching of the Y variables in file B on the marginal distributions is visualized in Figure 5.26. From the imputed data we calculate the statistics presented in Table 5.43 concerning the imputation of variables Y in file B. The qqplots tend to show rather straight lines. Comparing the moments of the distributions in the imputed and the original data sets supports the visual impression. Obviously MICE preserves the marginal distributions well. Even within the imputed variables Y only few differences in the parameter estimates are calculated.

Variable	% Zeros		Mean		Standard error	
X_i	p_i^{org}	$\frac{1}{5}\sum p_i^{(k)imp}$	$\overline{X}_i^{org}$	$\frac{1}{5}\sum \overline{X}_i^{(k)imp}$	S_i^{org}	$\frac{1}{5}\sum S_i^{(k)imp}$
X_{15}	0.31	0.31	0.09	0.09	0.14	0.13
X_{16}	0.20	0.20	0.17	0.16	0.21	0.20
X_{17}	0.12	0.13	0.28	0.27	0.26	0.26
X_{18}	0.14	0.15	0.18	0.18	0.19	0.19
X_{19}	0.05	0.06	0.42	0.41	0.28	0.28
X_{20}	0.02	0.01	0.43	0.42	0.25	0.24
X_{21}	0.01	0.01	0.49	0.49	0.25	0.24

Table 5.42. MICE: marginal distributions of imputed variables X_{15}, X_{16}, and X_{17}

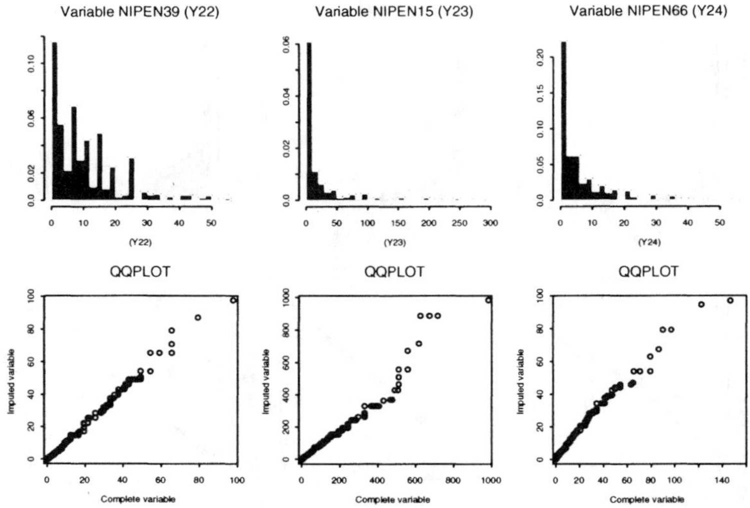

Figure 5.26. MICE: histogram and qqplot of imputed variables Y_{22}, Y_{23}, and Y_{24}

Goodness-of-Fit

Remember that MICE is not designed to use a specific prior distribution for the model parameters. The underlying Bayesian regression models always assume the usual noninformative prior distributions; see van Buuren and Oudshoorn (2000). However, the imputations are based on the regression of the variable to be imputed in the current step on all other variables whereas some are imputed themselves. In the simplified setting of our simulation study in Chapter 4 MICE establishes conditional independence of the variables never jointly observed given the common variables.

Variable	% Zeros		Mean		Standard error	
Y_i	p_i^{org}	$\frac{1}{5}\sum p_i^{(k)imp}$	$\overline{Y}_i^{org}$	$\frac{1}{5}\sum \overline{Y}_i^{(k)imp}$	S_i^{org}	$\frac{1}{5}\sum S_i^{(k)imp}$
Y_{22}	0.07	0.07	10.72	10.95	10.83	11.03
Y_{23}	0.22	0.21	33.66	33.04	77.78	77.63
Y_{24}	0.27	0.27	6.22	6.04	9.88	9.17
Y_{25}	0.05	0.06	8.22	8.07	8.56	8.30
Y_{26}	0.37	0.37	1.38	1.28	2.48	2.29
Y_{27}	0.09	0.09	8.95	8.95	8.14	8.16
Y_{28}	0.38	0.39	1.80	1.80	3.40	3.37
Y_{29}	0.32	0.34	0.04	0.05	0.27	0.30
Y_{30}	0.32	0.32	1.55	1.54	3.83	3.77

Table 5.43. MICE: marginal distributions of imputed variables Y_{22}, Y_{23}, and Y_{24}

File	Equal % zeros			KS			Wilcoxon			Paired t		
A	No a	a_{300}	a_{600}	No a	a_{300}	a_{600}	No a	a_{300}	a_{600}	No a	a_{300}	a_{600}
X_{15}	3	4	5	3	4	4	3	5	5	4	5	4
X_{16}	5	5	5	1	2	4	1	1	3	1	1	4
X_{17}	4	5	5	2	2	0	2	1	0	1	2	0
X_{18}	4	5	5	0	0	3	1	0	2	1	0	2
X_{19}	1	5	5	2	3	4	4	5	4	4	5	4
X_{20}	5	5	5	4	4	4	3	4	5	5	4	5
X_{21}	5	3	5	5	4	4	5	4	3	5	3	4
$\sum$	27	32	35	17	19	23	19	20	22	21	20	23

File	Equal % zeros			KS			Wilcoxon			Paired t		
B	No a	a_{300}	a_{600}	No a	a_{300}	a_{600}	No a	a_{300}	a_{600}	No a	a_{300}	a_{600}
Y_{22}	5	5	4	5	5	5	4	5	5	4	3	5
Y_{23}	5	4	5	5	5	5	5	5	5	5	5	4
Y_{24}	5	3	5	5	4	5	5	3	5	5	3	5
Y_{25}	5	4	5	4	3	5	5	5	5	4	5	3
Y_{26}	5	5	5	5	5	5	3	3	5	3	3	5
Y_{27}	4	5	2	5	4	4	4	2	4	4	2	4
Y_{28}	5	4	5	5	4	5	4	5	5	4	5	5
Y_{29}	5	5	5	5	5	5	4	5	5	4	5	5
Y_{30}	5	5	5	5	5	5	5	5	4	5	5	4
$\sum$	44	40	41	44	40	44	39	38	43	38	36	40

Table 5.44. MICE: comparison of marginal distributions concerning survey data Y

(1) Comparison of marginal distributions

Testing the equality of parameters estimated from the imputed and the original data sets yields the best agreement we have found throughout this simulation study. The results are given in Table 5.44. Obviously the imputation procedure MICE allows a good reproduction of the proportions of zeros. For the X variables the Kolmogorov–Smirnov test does not reject the hypothesis of equal distributions in the majority of cases. When the distributions of the imputed Y variables are compared with the original distributions then this hypothesis is only rejected in some few cases. This might be due to the imputation of observed values rather than random draws. The preservation of the continuous part of the X variables is less accurate than with NIBAS or NORM, but the distributions of the Y variables are sustained best by MICE.

(2) Bivariate associations

Concerning the bivariate association of each pair of the X and Y variables we achieve the results presented in Table 5.45. The majority of the bivariate associations are regarded as zero. Less than half of the positive as well as negative correlations are reported to be the same in the original and the imputed data. The number of false negatives is higher than it was with the other techniques. The reproduction of the bivariate association is improved by using auxiliary information: the larger the auxiliary file the better.

(3) Conditional independence

Inference about the bivariate conditional independence of Y and X given the common variables Z is summarized in Table 5.46. Without the auxil-

	Based on Spearman's r_S					
	Impute X in A			Impute Y in B		
Inference	No a	a_{300}	a_{600}	No a	a_{300}	a_{600}
Correct	23	24	33	13	19	30
False pos.	0	0	0	0	0	0
False neg.	40	39	30	50	44	33
Contradiction	0	0	0	0	0	0

	Based on Pearson's r_P					
	Impute X in A			Impute Y in B		
Inference	No a	a_{300}	a_{600}	No a	a_{300}	a_{600}
Correct	21	26	41	23	28	34
False pos.	0	0	0	0	0	0
False neg.	42	37	22	40	35	29
Contradiction	0	0	0	0	0	0

Table 5.45. MICE: inference about bivariate associations of survey data X and Y

Inference based on $\widehat{\beta}_{YX.Z}$	Impute X in A No a	a_{300}	a_{600}	Impute Y in B No a	a_{300}	a_{600}
Correct	37	50	39	47	36	43
False pos.	0	0	3	0	0	0
False neg.	26	13	21	16	27	20
Contradiction	0	0	0	0	0	0

Table 5.46. MICE: inference about conditional independence of survey data X and Y given $Z = z$

iary file and with the small auxiliary file, all of the conditional associations are regarded as being zero. Obviously this imputation procedure tends to establish conditional independence as expected with traditional imputation techniques. These results exactly coincide with our simulation study presented in Section 4.8. There we showed that the multiple imputation procedure MICE always establishes conditional independence due to the identification problem because no prior information can be used at present. Thus, the results produced by MICE are similar to those produced by NIBAS under the prior of zero. Applying the multiple imputation test procedures additionally widens the confidence intervals tending to report even more conditional independence.

Although many times conditional independence is reported for two variables X_i and Y_j given $Z = z$ for $i = 15, 16, \ldots, 21$, $j = 22, 23, \ldots, 30$, the null hypothesis concerning independence of the blocks of the variables X and Y given Z is always rejected; see Table 5.47. We find the result shown in Table 5.47 quite remarkable when no auxiliary information is available. Table 5.45 shows a tendency of MICE to impute the variables towards zero (unconditional) correlation of X and Y. Table 5.46 obviously displays that every two variables X and Y given $Z = z$ tend to be conditionally uncorrelated after the match although conditional noncorrelation is not reported when the whole block of variables X and Y given $Z = z$ is investigated. The conservative multiple imputation test procedures concerning the bivariate associations do not report much association but in each imputed data set the hypothesis of X and Y being conditionally uncorrelated given $Z = z$ is rejected by the usual test procedure. It seems that MICE tends to impute X as well as Y with only a small amount of association.

Statistic	Impute X in A No a	a_{300}	a_{600}	Impute Y in B No a	a_{300}	a_{600}
Mean dev.	650.07	561.72	376.37	550.87	405.89	297.63
Counts	0	0	0	0	0	0

Table 5.47. MICE: inference about the mean deviance of survey data X and Y given $Z = z$

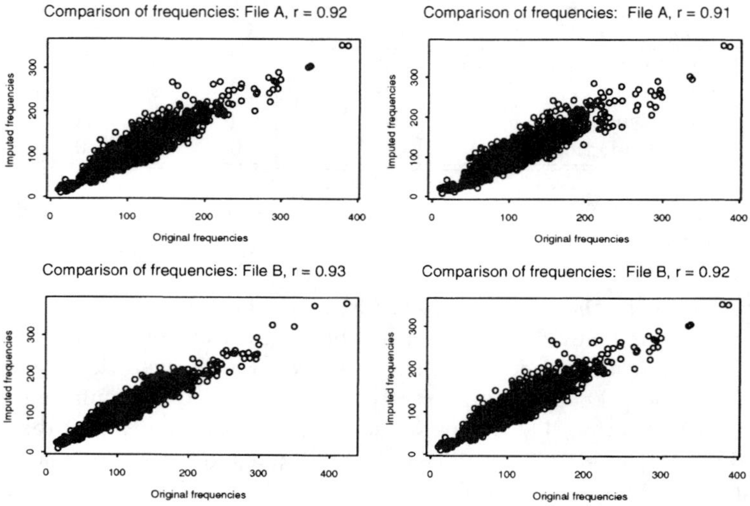

Figure 5.27. MICE: best and worst cross-tabulation fit

(4) Cross-tabulation fit

MICE produces the best cross-tabulation fit when the correlation coefficient is computed without correcting for a marginal fitting. This might be due to the fact that only observed values are imputed. Figure 5.27 displays the worst and the best cross-tabulation fit from the five imputed data sets when no auxiliary file is used. The mean correlation coefficients are given in Table 5.48. The null hypothesis that every two pairs of categorized bivariate distributions of X and Y in the imputed and the original data come from the same population is not often rejected; see Table 5.49. Like the univariate distributions the bivariate categorized distributions are also sustained quite well by MICE. These are the best results for the multiple

	Impute X in A			Impute Y in B		
Statistic	No a	a_{300}	a_{600}	No a	a_{300}	a_{600}
$\frac{1}{m}\sum r^{(k)}$	0.9119	0.9021	0.9231	0.9302	0.9300	0.9287
$\frac{1}{m}\sum r_{KW}^{(k)}$	0.6585	0.6503	0.6835	0.6990	0.6822	0.7354

Table 5.48. MICE: cross-tabulation fit of survey data

	Impute X in A			Impute Y in B		
Statistic	No a	a_{300}	a_{600}	No a	a_{300}	a_{600}
$[\min, \max]$	[29, 42]	[26, 39]	[43, 54]	[45, 56]	[50, 57]	[53, 60]

Table 5.49. MICE: inference about cross-tabulation fit of survey data

Inference based on regression	Impute X in A			Impute Y in B		
	No a	a_{300}	a_{600}	No a	a_{300}	a_{600}
Correct	178	182	177	162	164	169
False pos.	2	5	5	2	6	0
False neg.	18	11	16	34	28	29
Contradiction	0	0	0	0	0	0

Table 5.50. MICE: counting linear regression inferences based on survey data

imputation procedures when based on the survey data. The auxiliary files seem to be very helpful in stabilizing the reproduction of the bivariate categorized distributions.

(5) Linear regression inference
The linear models are computed regressing every Y variable on all X variables and the common variables Z. Notice that the same regression model is used to generate the multiple imputations. The results counting the conclusions made at the 5% significance level are listed in Table 5.50. The regression-based inference is rather similar to the noniterative multivariate procedure NIBAS due to the conditional independence assumption used. Also the auxiliary files do not improve inference much.

Using Generated Data

Taking the generated data from $(Z, X, Y) \sim N_{30}(\widehat{\mu}, \Sigma^{\star})$ we calculate all the tests again.

(1) Comparison of marginal distributions
The results of the tests comparing the distributions of the imputed and the original data are shown in Tables 5.51 and 5.52. In most of the cases the auxiliary information does not improve the imputation procedure

File	KS			Wilcoxon			Paired t		
A	No a	a_{300}	a_{600}	No a	a_{300}	a_{600}	No a	a_{300}	a_{600}
X_{15}	5	5	5	5	5	5	5	5	4
X_{16}	5	5	5	5	5	5	5	5	5
X_{17}	5	5	5	5	5	5	5	5	5
X_{18}	5	4	4	5	4	3	5	4	4
X_{19}	5	4	4	4	4	3	4	4	3
X_{20}	5	4	5	5	4	5	5	4	5
X_{21}	1	1	1	3	2	4	4	2	4
$\sum$	31	28	29	32	29	30	33	29	30

Table 5.51. MICE: comparison of marginal distributions concerning generated data X

File	KS			Wilcoxon			Paired t		
B	No a	a_{300}	a_{600}	No a	a_{300}	a_{600}	No a	a_{300}	a_{600}
Y_{22}	4	4	5	4	5	5	4	5	5
Y_{23}	5	4	5	4	3	3	4	2	3
Y_{24}	5	5	4	5	5	4	5	5	4
Y_{25}	5	5	4	5	5	4	5	5	4
Y_{26}	5	5	5	4	5	4	4	3	4
Y_{27}	4	2	5	5	4	5	5	5	5
Y_{28}	4	4	5	5	5	5	5	4	5
Y_{29}	2	5	5	5	5	5	5	5	5
Y_{30}	5	4	5	5	3	4	5	3	4
$\sum$	39	38	43	42	40	39	42	37	39

Table 5.52. MICE: comparison of marginal distributions concerning generated data Y

concerning the preservation of the marginal distributions. The marginal distributions are preserved well; in general, the distributions are reported to be the same more often by the Wilcoxon signed rank test and the two-sample paired t-test than the other imputation procedures.

(2) Bivariate associations

The preservation of the unconditional bivariate association of X and Y again gives evidence that MICE establishes (unconditional) independence between X and Y when no auxiliary information is available; see Table 5.53. Without the auxiliary file the bivariate associations are poorly reproduced. Obviously MICE tends to impute the variables not jointly observed toward

	Based on Spearman's r_S					
	Impute X in A			Impute Y in B		
Inference	No a	a_{300}	a_{600}	No a	a_{300}	a_{600}
Correct	6	35	52	8	40	48
False pos.	0	0	0	1	0	0
False neg.	57	28	11	54	23	15
Contradiction	0	0	0	0	0	0

	Based on Pearson's r_P					
	Impute X in A			Impute Y in B		
Inference	No a	a_{300}	a_{600}	No a	a_{300}	a_{600}
Correct	7	35	52	7	37	49
False pos.	0	0	0	1	0	0
False neg.	56	28	11	55	26	14
Contradiction	0	0	0	0	0	0

Table 5.53. MICE: inference about bivariate associations of generated data X and Y

Inference based on $\widehat{\beta}_{YX.Z}$	Impute X in A No a	a_{300}	a_{600}	Impute Y in B No a	a_{300}	a_{600}
Correct	9	29	46	9	32	44
False pos.	0	1	1	0	0	1
False neg.	54	33	16	54	31	18
Contradiction	0	0	0	0	0	0

Table 5.54. MICE: inference about conditional independence of generated data X and Y given $Z = z$

independence. Most of the associations in the imputed data are treated as being zero. If auxiliary data are available from X, Y and Z, information about their association is better exploited when the auxiliary data sets are larger.

(3) Conditional independence
In addition to independence between Y_j and X_i, conditional independence of Y_j and X_i given the common variables Z for $i = 15, 16, \ldots, 21$, $j = 22, 23, \ldots, 30$ is also established by MICE according to Table 5.54.

Although we take a data set with only few variables X and Y being conditionally independent now all combinations drift to conditional independence. These results are quite similar to those produced by NIBAS due to the prior set to zero. Auxiliary information again improves the number of correct specified inferences although NIBAS makes better use of it.

Nevertheless, conditional independence is only reported for the bivariate distributions and not for the joint distributions of X and Y given $Z = z$; see Table 5.55. The results shown in Table 5.55 are similar to those based on the survey data. Without auxiliary information the multiple imputation test procedures do not report any bivariate association between each X and Y variable although there seems to be some association among the blocks of variables X and Y.

(4) Cross-tabulation fit
The cross-tabulation fit based on the categorized bivariate distributions is the worst of all; see Table 5.56.

The fit is no longer as good as it was with the survey data and with the other multiple imputation procedures. Only a larger auxiliary file improves

Statistic	Impute X in A No a	a_{300}	a_{600}	Impute Y in B No a	a_{300}	a_{600}
Mean dev.	1764.62	3023.06	6563.39	1747.69	3297.34	6940.69
Counts	0	0	0	0	0	0

Table 5.55. MICE: inference about the mean deviance of generated data X and Y given $Z = z$

Statistic	Impute X in A			Impute Y in B		
	No a	a_{300}	a_{600}	No a	a_{300}	a_{600}
$\frac{1}{m}\sum r^{(k)}$	0.2839	0.6818	0.7843	0.2568	0.6908	0.7964
$\frac{1}{m}\sum r_{KW}^{(k)}$	0.3138	0.7385	0.8235	0.3043	0.7455	0.8303

Table 5.56. MICE: cross-tabulation fit of generated data

Statistic	Impute X in A			Impute Y in B		
	No a	a_{300}	a_{600}	No a	a_{300}	a_{600}
[min, max]	[26, 38]	[42, 53]	[61, 63]	[25, 31]	[50, 51]	[60, 63]

Table 5.57. MICE: inference about cross-tabulation fit of generated data

the fit considerably. Even the hypothesis about the categorized bivariate distributions coming from the same population is more often rejected here. The results counting the numbers of not rejecting the hypothesis of homogeneity out of 63 possible combinations for each of the imputed data sets are listed in Table 5.57. Similar to our results in the simulation study we find a better exploration of the small auxiliary data sets by the other imputation techniques.

(5) Linear regression inference

The results based on the 198 regression coefficients from the regression of each Y variable on the X and Z variables are given in Table 5.58.

Similar to the imputations generated by NIBAS the conditional independence that is established when no auxiliary information is available leads to a high amount of false negatives. As in the previous procedure inference is improved by using auxiliary information although the other procedures are obviously exploiting this additional information better.

Summary

Overall we get a good reproduction especially of the marginal distributions by MICE. When the survey data are used then imputing observed values rather than randomly generated values seems to improve sustaining the marginal distributions considerably. Due to the missing possibility

Inference based on regression	Impute X in A			Impute Y in B		
	No a	a_{300}	a_{600}	No a	a_{300}	a_{600}
Correct	94	124	152	91	115	139
False pos.	6	3	4	2	1	2
False neg.	96	70	42	103	82	57
Contradiction	2	1	0	2	0	0

Table 5.58. MICE: counting linear regression inferences based on generated data

of applying any suitable prior to overcome the problem of inestimability obviously a mixture of conditional and unconditional independence is established by MICE after the match. Since the amount of conditionally independent variables X and Y given $Z = z$ is already high within the survey data, good results concerning the cross-tabulation fit are produced by MICE. The auxiliary files are of some help, since they improve the reproduction of the correlation structure as was shown clearly with the generated data set.

Making further investigations of the imputations produced by MICE we realize that the between-imputation variance of the multiple imputation estimates in general is a bit smaller than the between-imputation variance derived from the data augmentation imputations but larger than the between-imputation variance produced by the noniterative multivariate model.

5.6 Concluding Remarks

In general and in accordance with the multiple imputation principle, we realize that the multiple imputation test procedures tend to be conservative based on the matched data. Compared with the usual complete-data statistics calculated using the original data sets the multiple imputation test procedures reject the null hypotheses less often. This is to be expected because the matched data are obtained from the incomplete data containing less information about the association of the variables never jointly observed. On the other hand, the multiple imputation test procedures additionally widen the confidence intervals to account for the imputation's uncertainty, therefore these tests are even more conservative.

If no auxiliary information is available prior information can most efficiently be used by the new noniterative multivariate procedure NIBAS assuming, for example, conditional independence of the variables not jointly observed. Different prior information may be used to display the sensitivity of the resulting estimates and the explanatory power of the common variables. The data augmentation procedure NORM introduces prior information via some hyperparameters that tend to establish (unconditional) independence of the variables never jointly observed. Finally, MICE does not allow the explicit use of prior information. The imputations generated by MICE seem to introduce conditional as well as unconditional independence among the specific variables not jointly observed. For NIBAS and NORM it is shown that the marginal distributions are better preserved the better the data model assumptions of normality are achieved. Thus, transforming the data suitably is important to achieve the fourth level of validity.

$\frac{1}{63}\sum B_k$	Survey data, no a NIBAS	NORM	MICE
Spearman's r_S	0.0003	0.0041	0.0011
Pearson's r_P	0.0003	0.0032	0.0017
CI $\widehat{\beta}_{YX.Z}$	8.8416	59.2087	29.4645

Table 5.59. Averaged between-imputation variances

Obviously auxiliary information does not improve the validity of the matching techniques concerning the preservation of the marginal distributions. They are well preserved in the majority of cases and best achieved by MICE. The third level of validity is better achieved by every technique if auxiliary information is used; here MICE seems to be less efficient.

Moreover, we found that NIBAS produces quite stable imputations; i.e., imputations generated by NIBAS yield smaller between-imputation variances than the other procedures. Table 5.59, for example, displays an overall mean of the between-imputation variances B_k averaged on the 63 possible combinations $k = 1, 2, \ldots, 63$. The differences are obvious.

Comparing other settings too, we find that most of the time the averaged between-imputation variance produced by MICE is a little bit smaller than the averaged between-imputation variance produced by NORM. It is always smaller with imputations from the noniterative multivariate procedure NIBAS. Thus, we conclude that one might stay with at least one single imputation produced by NIBAS, taking into account that the usual complete case analysis is less conservative than the multiple imputation test procedures would be.

Some features of the proposed imputation procedures are summarized in Figure 5.28.

It is shown that these Bayesian approaches as proposed here are easy to compute even with a larger data set; hence they might become interest-

Some properties	Procedure NIBAS	NORM	MICE
Speed	×	×	
Stochastic data model only for variables to be imputed	×		×
Model for continuous and discrete variables to be imputed			×
Application for missing data problems in general		×	×
Standalone software version available		×	
Non-iterative imputations	×		
Bayesianly proper multiple imputations	×	×	
Parameters drawn from the observed data posterior	×		
Model based on multivariate structure of the data	×	×	
Allows efficient use of third (complete) data sources	×	×	×
Allows efficient use of a prior	×	×	

Figure 5.28. Some comparative features of NIBAS, NORM, and MICE

ing alternatives to practitioners. Based on the results presented herein we finally draw some conclusions.

- The marginal distributions are already well preserved in the majority of cases and best achieved by MICE. When the data are suitably transformed, our fourth level of validity seems to be well achieved by all alternative techniques.
- NIBAS is the only routine available that explicitly allows us to control the conditional association of the variables not jointly observed via its prior distribution. We again find that NIBAS establishes conditional independence if it is assumed a priori. Thus, with different prior settings, we propose to use NIBAS for evaluation of the predictive power of the common variables.
- Whenever auxiliary data are available we find NORM very efficiently using them. Together with the standalone software NORM 2.02 or 2.03 we think it is a good alternative for practice. It is easy to handle and already provides many useful transformations, e.g., for skewed or count data.
- Whenever different types of variables have to be imputed and auxiliary data are available MICE can be easily applied although it might suffer from some theoretical disadvantages. The preservation of the marginal distribution seems to be very good especially with predictive mean matching.

6
Synopsis and Outlook

Fusion of databases may not be the best solution, but it might be the only practical solution as long as single-source information is unavailable, impractical or prohibitively expensive to obtain. *J.C. Adamek (1994)*

6.1 Synopsis

6.1.1 Summary

In Chapter 1 we point out that the statistical matching task may be viewed as a problem of nonresponse; more precisely, the missing information is regarded as missing at random because the missingness is induced by the study design of the separate samples. The missing data are due to unasked questions and the missingness mechanism is regarded as ignorable which in principle makes the application of conventional multiple imputation techniques obvious. However, contrary to the traditional missingness patterns, statistical matching is characterized by its identification problem. The association of the variables never jointly observed is unidentifiable and cannot be estimated by means of likelihood inference. Therefore prior information has to be embedded in the estimation process. Statistically matched files tend to display conditional independence between the variables only observed in separate files. We show that the validity of the traditional matching techniques concerning the preservation of the true association of the variables never jointly observed depends on the explanatory power of the common variables. Following an approach published by Rubin (1987) we propose the use of multiple imputation techniques using informative prior distributions to overcome the conditional independence assumption. By means of MI, sensitivity of the unconditional association of the (specific) variables not jointly observed can be displayed. In other words, different prior settings of conditional associations allow us to show the extent to which unconditional associations are determined by the common variables.

Throughout Chapter 2 a frequentist theory is derived concluding that the usual statistical matching techniques, relying on any sort of nearest neighbor matches, will preserve the distributions of the variables in the separate files but lead to conditional independence of the specific variables in the statistically matched file. Simulation studies are presented displaying no or only few influences of different marriage algorithms or distance functions on the assumption of conditional independence. Even matching based on propensity scores (i.e., when a rather coarse distance function is used) will imply conditionally independent variables in the statistically matched file with an additional distortion of the joint distribution of the specific and the common variables. Finally, the simulation studies indicate that the choice of suitable common variables is most important concerning the validity of a statistical matching technique.

Based on the marginal distributions of the variables in the separate files and the joint distributions of the variables not jointly observed, we propose four levels of validity that a matching technique may achieve. The levels are ordered and discussed in Chapter 2. The fourth level of validity requiring the preservation of the marginal distributions is postulated to be the minimum validity. Preserving the association of the specific variables is defined to be the third level. Sustaining the joint distribution leads to the second level, which is regarded to be the most important target of statistical matching. Assuming that the matched data often are used for any kind of individual analyses, the second level of validity is required to ensure the validity of these analyses. The first level of recovering the true values is of minor importance especially when continuous variables are matched.

Traditional approaches as they are currently applied in practice are the focus of the third chapter. Historically statistical matching was "invented" and extended in the early 1960s and 1970s in Germany and France to match print media information and television viewing behavior with purchasing information due to media planning needs. In the USA and Canada since the 19070s surveys have been matched by federal offices to achieve better comprehensive household income information. Unconstrained as well as constrained matching is explained in detail. Current techniques to merge data from different sources are reported such as the categorical constraint matching practiced by Statistics Canada, the topological concept used in Germany, and the multiple ascription approach first applied in the UK. Summarizing the results we find that the known techniques address the search of suitable common variables that already lead to conditional independence. However, all these traditional techniques suffer from the lack of a measure to investigate the quality of the matching process as well as the profit of prior information.

The disadvantages of the traditional approaches lead to the attempt to visualize the identification problem inherent in statistical matching by means of the multiple imputation approach. Therefore, in Chapter 4 the concepts of observed-data likelihood and observed-data posterior distributions are first discussed for the reader unfamiliar with the usual assumptions made for the imputation of missing data. The theorem of ignorability is described briefly and the missing data mechanism is identified as an ignorable one. Then we show why the well-known EM algorithm does not work in the case of matching. Those principles of Bayesian inference which are essential for understanding our multiple imputation procedure are described. For illustration purposes notation and necessary derivations are explained using simplified examples of statistical matching. Then the multiple imputation principle is introduced. In the following we discuss an imputation procedure with random residuals to perform MI. Next this approach is generalized to the non-iterative Bayesian multivariate linear regression model NIBAS, then the data augmentation algorithm NORM is described, and the procedure MICE based on regression switching algorithms is presented. Finally, a simulation study using a conventional multivariate normal model illustrates the advantages and disadvantages of the different techniques proposed. The estimation of the bounds of the unconditional association of the variables not jointly observed is shown. By means of the simulations study we realize that prior information can most easily be applied by NIBAS. MICE does not allow the use of prior information yet. With NORM prior information is introduced via some hyperparameters when the stand alone MS-Windows^{TM} version is used. Furthermore we discuss the gain of auxiliary information provided by a small but complete data source. In the simulation study the auxiliary file is most efficiently exploited by NORM and second by NIBAS, whereas MICE obviously needs larger auxiliary data sets.

In the fifth chapter an extensive empirical evaluation based on real survey data from a television measurement panel is presented to show the practical importance of the alternative techniques proposed. The data are provided from the television behavior panel run by the largest German market research company. Thus we had access to nearly 5500 observations with a variety of variables concerning the demographics of the panel members, their average TV viewing time, their (estimated) probabilities of viewing a specific channel, and their consuming behavior. The focus of this chapter and the study presented there is less to achieve estimated bounds of the unconditional association but to compare inferences based on the imputed data with inferences based on the actual complete data. Since the multiple imputation techniques are model-based much effort is taken to meet the model assumptions, especially those concerning normality. Transformations are presented to operate variables that are typically skewed, having a large amount of zeros. The survey data we had access to already contained common variables with high explanatory power; thus the amount of specific variables being conditionally uncorrelated is rather high. To investigate the properties of the different imputation techniques proposed we generate a data set with a similar structure to the survey data but a higher amount of conditional dependence. Hence we find that NIBAS establishes conditional independence if it is assumed a priori. NORM tends to produce unconditional independence and MICE a mixture of both not allowing us to use prior information in any way. The marginal distributions are already well preserved in the majority of cases and best preserved by MICE. Auxiliary information is helpful for improving the preservation of the correlation structure; here MICE seems to be less efficient. Principally better matching results are achieved compared to inference drawn from the original data the better the assumption of normality is met.

Over all, compared to the original data the significance tests based on the imputed data for most of the techniques were more likely to accept independence when there was association in the original data than to reject the hypotheses of independence when there was independence in the original data. Thus, the tests of association based on the imputed data are more conservative than those based on the complete data. This is due to the multiple imputation principle which accounts for the model uncertainty as well as the imputation uncertainty. In general, we find that NIBAS produces quite stable imputations; i.e., imputations generated by NIBAS yield smaller between-imputation variances than the other procedures. If multiply imputed data sets are not accepted, we conclude that one might stay with at least one single imputation produced by NIBAS, taking into account that the usual complete case analysis is less conservative than the multiple imputation test procedures would be. Notice that NIBAS is designed especially to meet the matching task: the parameters of the model are drawn from their observed data posterior distribution and the missing values are imputed according to their posterior predictive distribution. No further iterations are necessary. NIBAS is an, at least, Bayesianly proper imputation method. An informative prior distribution is only used for the conditional associations of the variables never jointly observed. For the parameters that can be estimated from separate samples the usual noninformative prior distributions are assumed. Hence the identification problem is overcome by the applied informative prior and no estimation problems will occur during the imputation process.

6.1.2 Available MI-Software for Statistical Matching

In addition to the new noniterative Bayesian routine NIBAS other routines and software packages are available that may also be used as alternative solutions for statistical matching. However, most of them do not allow the use of an informative prior distribution. The multiple imputation meth-

Program	Available at	Speciality
NORM, CAT, MIX, PAN	http://www.stat.psu.edu/~jls	S-PLUS library
Missing Data Library	Insightful Corporation	S-PLUS library
NORM 2.03	http://www.stat.psu.edu/~jls	standalone (WIN)
AMELIA	http://gking.harvard.edu/stats.shtml	standalone (DOS)
Hmisc	http://hesweb1.med.virginia.edu/biostat/s	S-PLUS library
IVEWARE	http://www.isr.umich.edu/src/smp/ive/	SAS procedure
MICE V1.0	http://www.multiple-imputation.com	S-PLUS/ R library
PROC MI	SAS Institute Inc.	SAS procedure
SOLAS 3.0	Statistical Solutions Ltd.	standalone (WIN)

Figure 6.1. Features of multiple imputation software

ods implemented usually are based on noninformative prior distributions. Based on the example of MICE we have discussed in Chapter 4 that with uninformative priors the imputation will result in a conditional independence solution. In the simulation study using the survey data in Chapter 5 the result is not that clear. Therefore, in the context of statistical matching these routines may be used efficiently whenever auxiliary data are available. An overview of the majority of currently available multiple imputation software packages and routines is given in Figure 6.1. Notice that this list surely is not complete but it is expected to cover the most recent techniques available.

Few techniques are available that allow the use of informative priors. Therefore, the noniterative routine NIBAS is implemented as it is especially suitable for statistical matching.

6.1.3 Conclusions

It is shown that the traditional approaches tend to establish conditional independence between the variables never jointly observed given the common variables: the better the explanatory power of the common variable, the better all interesting levels of validity are achieved by these techniques. Unfortunately, the assumption of conditional independence is untestable by the traditional approaches. By means of multiple imputation procedures we make prior information usable. Under different prior settings for the conditional association of the variables never jointly observed the missing data are multiply imputed. From these multiply imputed data sets finally upper and lower bounds may be estimated to achieve a range of the unconditional association. We propose to use this range as a quality measure for a reliable matching technique. Additionally, we have shown that auxiliary information can be used efficiently by the alternative Bayesian approaches.

However, in large-scale surveys with different types of variables the model assumptions may be difficult to fulfill. Also the computations of the remaining bandwidth of the unconditional association will be more cumbersome to calculate the higher the dimension of the data set is. Thus, future research may be addressed to the calculation of these bounds in large mixed data sets and to the search for good predicting common variables that ensure the assumption of conditional independence. Moreover, the identification problem inherent in statistical matching may be avoided by approaches based on the so-called multiple matrix sampling, see section 6.2.2.

6.2 Outlook

6.2.1 Searching for Conditional Independence

Throughout the previous chapters we have discussed the importance of the conditional independence assumption especially when traditional matching techniques are applied. A high explanatory power of the common variables is desirable even for the alternative approaches to narrow the bounds of the unconditional association. Further research may be directed to methods searching for conditional independence in high-dimensional data structures based on small auxiliary files.

Notice that it can either be difficult or even impossible to identify common variables ensuring the assumption of conditional independence especially in large data sets. If it is not possible to find common variables that ensure conditional independence between the variables not jointly observed, it may be possible to construct unobserved, so-called "latent" classes to ensure conditional independence within these classes. An interesting approach to statistical matching relying on latent class analysis was proposed by Kamakura and Wedel (1997). This approach is especially attractive because the variables are assumed to be (conditionally) independent within such discrete latent classes; see Everitt (1984). When the observed variables in the different samples follow continuous as well as discrete distributions a generalization may be used which was proposed by Bacher (2000) and Rässler and Fleischer (2000) combining latent class and latent profile analysis.

The results published by Kamakura and Wedel (1997) are rather encouraging but their approach is restricted to estimating bivariate cell frequencies of a statistically matched file. A first application of the generalized latent class approach to the matching task was given by Rässler and Fleischer (2000). This approach still suffers from computational limitations, though. Nevertheless it seems a challenging area for further research.

6.2.2 Avoiding Identification Problems

Another very promising approach addressing the problem of identification was published by Raghunathani and Grizzle (1995) as well as Gelman et al. (1998). Gelman et al. (1998) proposed a Bayesian model to match (to impute) information from more than one sample, whereas some variables have not been observed in all samples. Raghunathani and Grizzle (1995)

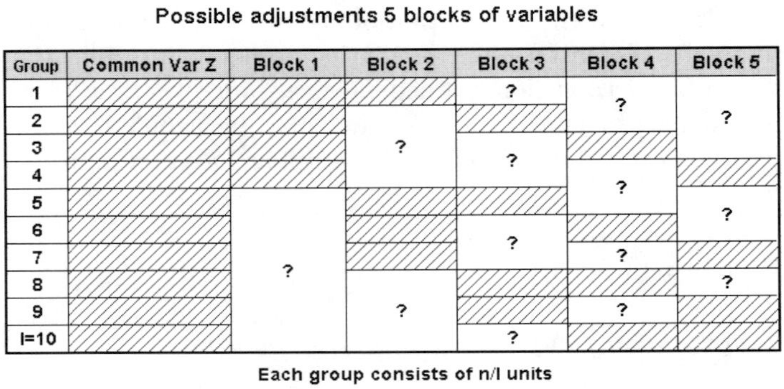

Figure 6.2. A split questionnaire survey design

proposed a "split questionnaire survey design" to arrange the variables to be surveyed in blocks that are not presented to all respondents of a survey. Thus a very typical pattern of missingness is generated that is best explained by an example. Consider, for instance, a questionnaire with $r = 110$ variables to be asked. Arbitrarily, let $k = 10$ variables be chosen to be answered by all respondents; the remaining $r - k = 100$ variables are split into s blocks with $(r - k)/s = 20$ variables each according to some correlation criteria. Every respondent is questioned about the (common) variables and some of the blocks. If analysis of the data set after imputation is restricted to second-order interactions then it is sufficient to generate $\binom{s}{2}$ $= l$ combinations of these s blocks. If n individuals are to be asked, then n/l individuals are chosen randomly to answer each combination. Figure 6.2 illustrates this situation. For example, Group one of the respondents is expected only to provide information about the common part Z and blocks one and two.

In the example illustrated in Figure 6.2 the response burden is reduced more than half for each respondent. Raghunathani and Grizzle (1995) proposed a multiple imputation procedure to draw the missing values from their posterior predictive distributions using a Gibbs sampling scheme. Notice that analysis of higher interactions requires that more blocks be given to a respondent. The combination of the blocks ensures that all parameters of the model are estimable from the observed data and, thus, no problems of identification occur.

The multiple imputation procedure proposed by Gelman et al. (1998) assumes a multiple imputation model allowing us to use covariates on the individual and the survey level. In principle a multivariate normal data model is assumed at the individual level with differing mean vectors concerning the survey levels and a common covariance matrix. The missingness pattern of the concatenated surveys ensures that all parameters of the model are estimable again.

The appealing idea of both approaches is that the missingness can be treated as missing at random due to unasked questions. Moreover, all combinations of variables are observed that are necessary to identify the parameters of the model. If the model parameters are identifiable no informative prior has to be used and the usual noninformative prior distributions may be applied without further hindrance.

Combining the ideas of these approaches with the matching task we propose to reorganize the usual surveys, for instance, the television measurement panel and the purchasing behavior panel, to create suitable blocks of variables. In Chapter 3 we have discussed that recent matching techniques especially emphasize the explanatory power of the specific common variables. For example, the purchasing behavior panel members already have been asked about their TV viewing habits. The television measurement panel members are asked additionally about their purchases. Reorganizing the panel surveys seems to be a further step, though. Hence, future research may be addressed to the creation of suitable split questionnaires to overcome the conditional independence assumption and the identification problems.

Appendix

Some Technicalities

Multivariate Normal Model Completely Observed

The data $U = (U_1, U_2, \ldots, U_r)$ are multivariate normally distributed with $U \sim N_r(\mu, \Sigma)$. Hence the joint distribution of U for any unit with observation $u = (u_1, u_2, \ldots, u_r)'$ is given by

$$f_{U|\Theta}(u|\theta) = (2\pi)^{-r/2}|\Sigma|^{-1/2} \exp\left\{-\frac{1}{2}(u-\mu)'\Sigma^{-1}(u-\mu)\right\}, \quad \theta = (\mu, \Sigma).$$

The covariance matrix Σ is an $r \times r$ matrix of the variances and covariances and μ is an r-dimensional vector of the means.

The likelihood of μ and Σ based on n independent observations u is

$$L(\mu, \Sigma; u) = (2\pi)^{-nr/2}|\Sigma|^{-n/2} \exp\left\{-\frac{1}{2}\sum_{i=1}^{n}(u_i-\mu)'\Sigma^{-1}(u_i-\mu)\right\},$$

with $u_i = (u_{i1}, u_{i2}, \ldots, u_{ir})'$. Let $S(\mu)$ denote the $r \times r$ symmetric matrix

$$S(\mu) = \sum_{i=1}^{n}(u_i-\mu)(u_i-\mu)';$$

then the likelihood function can be rewritten as

$$L(\mu, \Sigma; u) = (2\pi)^{-nr/2}|\Sigma|^{-n/2} \exp\left\{-\frac{1}{2}tr(\Sigma^{-1}S(\mu))\right\}.$$

For the prior distribution of the random variables $\Theta = (\mu, \Sigma)$, we assume first that μ and Σ are approximately independent with

$$f_{\mu,\Sigma}(\mu, \Sigma) \approx f_\mu(\mu) f_\Sigma(\Sigma)\,.$$

The typical setting addressing the missing data problem assumes either a normal distribution or a noninformative prior for μ. Since our interest is focused on the covariance matrix rather than the mean, we suppose that it is appropriate to take μ as locally uniform with

$$f_\mu(\mu) = c_{prior}, \quad c_{prior} \in \mathbb{R}^r;$$

i.e., this improper prior distribution of μ is assumed to be uniform over the r-dimensional space. Concerning the covariance of the variables never jointly observed we use either informative or improper prior information.

- As a suitable informative prior distribution of the $\frac{1}{2}r(r+1)$ distinct elements of the $r \times r$ (positive definite) covariance matrix Σ, usually the inverted-Wishart distribution is taken with some hyperparameters ν and Λ; i.e., $\Sigma \sim W_r^{-1}(\nu, \Lambda)$. Notice that $\nu > 0$ and Λ is an $r \times r$ positive definite symmetric matrix of fixed constants.
  This inverted-Wishart distribution is derived from the Wishart distribution of Σ^{-1}. In Bayesian analysis it is convenient to assume Σ^{-1} being Wishart-distributed with density function

$$f_{\Sigma^{-1}}(\Sigma^{-1}; \nu, \Lambda) = c^{-1}|\Sigma|^{-\nu/2+1} \exp\left\{-\frac{1}{2}tr(\Sigma^{-1}\Lambda^{-1})\right\},$$

  where c denotes a normalizing constant. It is

$$\begin{aligned} c &= \int |\Sigma|^{-\nu/2+1} \exp\left\{-\frac{1}{2}tr(\Sigma^{-1}\Lambda^{-1})\right\} d\Sigma \\ &= |\Lambda|^{(\nu+r-1)/2} 2^{r(\nu+r-1)/2} \Gamma_r\left(\frac{\nu+r-1}{2}\right), \end{aligned}$$

  with $\Gamma_r((\nu+r-1)/2)$ denoting the generalized gamma function; i.e.,

$$\begin{aligned} \Gamma_r\left(\frac{\nu+r-1}{2}\right) &= \left(\Gamma\left(\frac{1}{2}\right)\right)^{r(r-1)/2} \prod_{i=1}^{r} \Gamma\left(\frac{\nu+i-1}{2}\right), \quad v > 0, \\ \text{with } \Gamma(a) &= \int_0^\infty u^{a-1}e^{-u}du,\ a > 0; \end{aligned}$$

  see Box and Tiao (1992), p. 427.
  According to the rules of variable transformation we derive the distribution of Σ with

$$\begin{aligned} f_\Sigma(\Sigma; \nu, \Lambda) &= f_{\Sigma^{-1}}(\Sigma^{-1}; \nu, \Lambda)\left|\frac{\partial \Sigma^{-1}}{\partial \Sigma}\right| \\ &= c^{-1}|\Sigma|^{-(\nu/2+r)} \exp\left\{-\frac{1}{2}tr(\Sigma^{-1}\Lambda^{-1})\right\}, \end{aligned}$$

with c exactly as defined above; see Box and Tiao (1992), p. 460. Notice that if $\Sigma^{-1} \sim W_r(\nu, \Lambda)$ then $\Sigma \sim W_r^{-1}(\nu, \Lambda)$ which for simplicity is said to be inverted-Wishart. The hyperparameter ν is often called the degrees of freedom and Λ serves as a prior scale of Σ.

- As a noninformative prior for Σ,

$$f_\Sigma(\Sigma) \propto |\Sigma|^{-(r+1)/2}$$

may be used which arises, e.g., by applying Jeffreys' rule to Σ; i.e., $f_\Sigma \propto |I(\Sigma)|^{1/2}$.

If we multiply the likelihood by the improper prior $f_{\mu,\Sigma}(\mu, \Sigma) = c_{prior}|\Sigma|^{-(r+1)/2} \propto |\Sigma|^{-(r+1)/2}$, with c_{prior} arbitrarily set to one following Berger (1985), p. 84, we calculate the posterior distribution of μ and Σ given the data $U = u$,

$$\begin{aligned} f_{\mu,\Sigma|U}(\mu, \Sigma|u) &= c^{-1}L(\mu, \Sigma; u)f_{\mu,\Sigma}(\mu, \Sigma) \\ &= c^{-1}(2\pi)^{-nr/2}|\Sigma|^{-(n+r+1)/2} \exp\left\{-\frac{1}{2}tr(\Sigma^{-1}S(\mu))\right\}, \end{aligned}$$

according to Bayes' rule (4.10). Sometimes it is more convenient to work with Σ^{-1}, thus the posterior distribution of μ and Σ^{-1} given the data $U = u$ is $f_{\mu,\Sigma^{-1}|U}(\mu, \Sigma^{-1}|u) = f_{\mu,\Sigma|U}(\mu, \Sigma|u)\left|\partial\Sigma/\partial\Sigma^{-1}\right|$ with

$$\begin{aligned} f_{\mu,\Sigma^{-1}|U} &= c^{-1}(2\pi)^{-nr/2}|\Sigma|^{-(n+r+1)/2} \exp\left\{-\frac{1}{2}tr(\Sigma^{-1}S(\mu))\right\}|\Sigma|^{r+1} \\ &= c^{-1}(2\pi)^{-nr/2}|\Sigma^{-1}|^{(n-r-1)/2} \\ &\quad \exp\left\{-\frac{1}{2}tr\left((\Sigma/n)^{-1}(\mu-\overline{u})(\mu-\overline{u})'\right)\right\} \\ &\quad \exp\left\{-\frac{1}{2}tr(\Sigma^{-1}S(\overline{u}))\right\}, \end{aligned}$$

making use of partitioning

$$S(\mu) = n(\mu-\overline{u})(\mu-\overline{u})' + S(\overline{u}) \text{ with } S(\overline{u}) = \sum_{i=1}^{n}(u_i-\overline{u})(u_i-\overline{u})'.$$

$S(\overline{u})$ is proportional to the sample covariance matrix, and $\overline{u}$ is the sample mean vector $\overline{u} = (\overline{u}_1, \overline{u}_2, \ldots, \overline{u}_r)'$, $\overline{u}_j = \frac{1}{n}\sum_{i=1}^{n} u_{ij}$ for $j = 1, 2, \ldots, r$. The normalizing constant c is given by integration (see Box and Tiao (1992), p. 145) over the parameter spaces of μ and Σ^{-1},

$$\begin{aligned} c &= (2\pi)^{-nr/2}\int|\Sigma^{-1}|^{(n-r-1)/2}\exp\left\{-\frac{1}{2}tr(\Sigma^{-1}S(\overline{u}))\right\} \\ &\quad \underbrace{\int \exp\left\{-\frac{1}{2}n(\mu-\overline{u})'\Sigma^{-1}(\mu-\overline{u})\right\}d\mu}_{=(2\pi)^{r/2}|\Sigma/n|^{1/2}}\; d\Sigma^{-1} \end{aligned}$$

$$= (2\pi)^{r(1-n)/2} n^{-1/2} \int |\Sigma^{-1}|^{(n-r)/2-1} \exp\left\{-\frac{1}{2} tr(\Sigma^{-1} S(\overline{u}))\right\} d\Sigma^{-1}$$

$$= (2\pi)^{r(1-n)/2} n^{-1/2} S(\overline{u})^{-(n-1)/2} 2^{r(n-1)/2} \Gamma\left(\frac{n-1}{2}\right).$$

Notice that the last integral is equivalent to the normalizing constant of a Wishart distribution with $n - r$ degrees of freedom.

The exact posterior is derived with

$$f_{\mu,\Sigma^{-1}|U} =$$

$$\underbrace{(2\pi)^{-r/2} |(\Sigma/n)|^{-1/2} \exp\left\{-\frac{1}{2}(\mu - \overline{u})'(\Sigma/n)^{-1}(\mu - \overline{u})\right\}}_{(1)}$$

$$\underbrace{|\Sigma^{-1}|^{(n-r)/2-1} \exp\left\{-\frac{1}{2} tr(\Sigma^{-1} S(\overline{u}))\right\} S(\overline{u})^{(n-1)/2} 2^{-r(n-1)/2} \Gamma\left(\frac{n-1}{2}\right)^{-1}.}_{(2)}$$

The part of the posterior denoted by (1) is a multivariate normal density $f_{\mu|\Sigma,U}$ of μ given Σ and the data $U = u$; (2) denotes $f_{\Sigma^{-1}|U}$ which is a Wishart density with parameter $n - r$ and $S(\overline{u})^{-1}$. Thus, under the improper prior the complete-data posterior is given by

$$\Sigma|u \sim W_r^{-1}\left(n - r, S(\overline{u})^{-1}\right) \text{ and}$$
$$\mu|\Sigma, u \sim N_r\left(\overline{u}, \frac{1}{n}\Sigma\right).$$

When we apply the informative prior of Σ and the noninformative prior of μ and take into account the data model, the calculations to get $f_{\mu,\Sigma|U} = f_{\mu|\Sigma,U} f_{\Sigma|U}$ are similar to those presented above. After some algebra we achieve the complete-data posterior distribution $f_{\mu,\Sigma|U} = c^{-1} f_\mu f_\Sigma L(\mu, \Sigma; u)$ for μ and Σ given the data with

$$\Sigma|u \sim W_r^{-1}(\nu + n - 1, \Lambda'), \quad \Lambda' = (\Lambda^{-1} + S(\overline{u}))^{-1} \text{ and}$$
$$\mu|\Sigma, u \sim N_r\left(\overline{u}, \frac{1}{n}\Sigma\right)$$

for some hyperparameters ν and Λ.

Normally Distributed Data Not Jointly Observed

Consider the case when there are only two variables X and Y never jointly observed. For the purpose of demonstration we do not take into account a third common variable Z here. This example has previously been addressed by Rubin (1987), pp. 187–188, and serves as the originator for our noniterative multivariate imputation procedure presented in Chapter 4.

The following data model is assumed.

$$Y, X | \mu_Y, \mu_X, \sigma_Y^2, \sigma_X^2, \rho \sim N_2(\mu, \Sigma)$$

$$\text{with}\mu = \begin{pmatrix} \mu_Y \\ \mu_X \end{pmatrix}, \Sigma = \begin{pmatrix} \sigma_Y^2 & \rho\sigma_Y\sigma_X \\ \rho\sigma_Y\sigma_X & \sigma_X^2 \end{pmatrix}.$$

Given n independent observations

$$\begin{aligned}(x, y) &= ((NA, y_1), (NA, y_2), \ldots, (NA, y_{n_A}), \\ &\quad (x_{n_A+1}, NA), (x_{n_A+2}, NA), \ldots, (x_n, NA)),\end{aligned}$$

i.e., $y = (y_1, y_2, \ldots, y_{n_A})$ from file A and $x = (x_1, x_2, \ldots, x_{n_B})$ from file B, the observed-data likelihood can be written as

$$\begin{aligned} L(\mu, \Sigma; x, y) &= \prod_{i=1}^{n_A} f_Y(y_i; \mu_Y, \sigma_Y^2) \prod_{i=n_A+1}^{n} f_X(x_i; \mu_X, \sigma_X^2) \\ &= (2\pi)^{-n_A/2} (\sigma_Y^2)^{-n_A/2} \exp\left\{ -\frac{1}{2\sigma_Y^2} \sum_{i=1}^{n_A} (y_i - \mu_Y)^2 \right\} \\ &\quad (2\pi)^{-n_B/2} (\sigma_X^2)^{-n_B/2} \exp\left\{ -\frac{1}{2\sigma_X^2} \sum_{i=n_A+1}^{n} (x_i - \mu_X)^2 \right\} \\ &= L(\mu_X, \sigma_X^2; x) L(\mu_Y, \sigma_Y^2; y). \end{aligned}$$

As a prior distribution $f_{\mu,\Sigma}$ let us choose a noninformative prior with μ and Σ being approximately independent $f_{\mu,\Sigma} \approx f_\mu f_\Sigma = f_\mu f_{\sigma_X^2, \sigma_Y^2|\rho} f_\rho$. A suitable noninformative prior is then given by

$$f_{\mu,\Sigma} \approx f_\mu f_{\sigma_X^2|\rho} f_{\sigma_Y^2|\rho} f_\rho = c_{prior} f_{\sigma_X^2|\rho} f_{\sigma_Y^2|\rho} f_\rho = c_{prior} \sigma_X^{-2} \sigma_Y^{-2} f_\rho.$$

According to Bayes' theorem (4.10) the joint posterior distribution $f_{\mu,\Sigma|X,Y}(\mu, \Sigma|x, y)$ equates $c^{-1}\ L(\mu, \Sigma; x, y)\sigma_X^{-2}\sigma_Y^{-2} f_\rho$ again setting the constant c_{prior} to one. We get

$$\begin{aligned} & f_{\mu,\Sigma|X,Y}(\mu, \Sigma|x, y) \\ = \ & c^{-1} (2\pi)^{-n_A/2} (\sigma_Y^2)^{-n_A/2-1} \exp\left\{ -\frac{1}{2\sigma_Y^2} \sum_{i=1}^{n_A} (y_i - \mu_Y)^2 \right\} \\ & (2\pi)^{-n_B/2} (\sigma_X^2)^{-n_B/2-1} \exp\left\{ -\frac{1}{2\sigma_X^2} \sum_{i=n_A+1}^{n} (x_i - \mu_X)^2 \right\} f_\rho, \\ = \ & \underbrace{c_Y^{-1} L(\mu_Y, \sigma_Y^2; y) f_{\sigma_Y^2|\rho}(\sigma_Y^2|\rho)}_{(1)} \underbrace{c_X^{-1} L(\mu_X, \sigma_X^2; x) f_{\sigma_X^2|\rho}(\sigma_X^2|\rho)}_{(2)} f_\rho(\rho). \end{aligned}$$

The normalizing constant c is calculated by

$$c = \underbrace{\int\int L(\mu_Y, \sigma_Y^2; y) f_{\sigma_Y^2|\rho}(\sigma_Y^2|\rho) d\mu_Y d\sigma_Y^2}_{=c_Y}$$

$$\underbrace{\int\int L(\mu_X, \sigma_X^2; x) f_{\sigma_X^2|\rho}(\sigma_X^2|\rho) d\mu_X d\sigma_X^2}_{=c_X} \underbrace{\int f_\rho(\rho) d\rho}_{=1} .$$

Using suitable integral formulas, for example, c_Y is given by

$$\begin{aligned}
c_Y &= (2\pi)^{-n_A/2} \int\int (\sigma_Y^2)^{-n_A/2-1} \exp\left\{-\frac{n_A(\mu_Y - \overline{y})^2}{2\sigma_Y^2}\right\} \\
&\quad \exp\left\{-\frac{1}{2\sigma_Y^2} \sum_{i=1}^{n_A} (y_i - \overline{y})^2\right\} d\mu_Y d\sigma_Y^2 \\
&= (2\pi)^{-n_A/2} \int (\sigma_Y^2)^{-n_A/2-1} \exp\left\{-\frac{1}{2\sigma_Y^2} \sum_{i=1}^{n_A} (y_i - \overline{y})^2\right\} \\
&\quad \underbrace{\int \exp\left\{-\frac{n_A(\mu_Y - \overline{y})^2}{2\sigma_Y^2}\right\} d\mu_Y}_{=\sqrt{2\pi\sigma_Y^2/n_A}} d\sigma_Y^2 \\
&= (2\pi)^{-n_A/2+1/2} n_A^{-1/2} \left(\frac{1}{2}\sum_{i=1}^{n_A} (y_i - \overline{y})^2\right)^{-(n_A-1)/2} \Gamma\left(\frac{n_A - 1}{2}\right).
\end{aligned}$$

Here we derive

$$(1) = \underbrace{(2\pi)^{-1/2} \left(\frac{\sigma_Y^2}{n_A}\right)^{-1/2} \exp\left\{-\frac{1}{2}\left(\frac{\mu_Y - \overline{y})}{\sigma_Y/\sqrt{n_A}}\right)^2\right\}}_{\Rightarrow \mu_Y|\sigma_Y^2, y \sim N(\overline{u}, \sigma_Y^2/n_A)}$$

$$\underbrace{\left(\Gamma\left(\frac{n_A - 1}{2}\right) 2^{(n_A-1)/2}\right)^{-1} (s_Y^2)^{(n_A-1)/2} (\sigma_Y^2)^{-((n_A-1)/2+1)} \exp\left\{-\frac{s_Y^2}{2\sigma_Y^2}\right\}}_{= f_{\sigma_Y^2|Y}(\sigma_Y^2|y)},$$

with $s_Y^2 = \sum_{i=1}^{n_A} (y_i - \overline{y})^2$. Notice that $f_{\sigma_Y^2|Y}(\sigma_Y^2|y) s_Y^2$ yields an inverted χ^2 distribution with $n_A - 1$ degrees of freedom. By the same argumentation we get $\sigma_Y^2|y \sim \chi_{(n_A-1)}^{-2} s_Y^2$ as well as the posterior distributions of $\mu_X|\sigma_X^2, x$ and $\sigma_X^2|x$ derived from (2). Moreover we find that the posterior distribution of ρ equals its prior distribution; i.e., $f_{\rho|U}(\rho|u) = f_\rho(\rho)$.

Summarizing the results derived above, we finally get the joint observed-data posterior $f_{\mu,\Sigma|X,Y}$ for μ and Σ given the data $Y = y$ and $X = x$ by means of their conditional and marginal observed-data posterior distributions,

$$\begin{aligned} f_{\mu,\Sigma|X,Y} &= f_{\mu_X,\sigma_X^2|\rho,X} f_{\mu_Y,\sigma_Y^2|\rho,Y} f_{\rho|X,Y} \text{ with} \\ \mu_Y|\sigma_Y^2, y &\sim N_1(\widehat{\mu}_Y, \sigma_Y^2/n_A), \\ \sigma_Y^2|y &\sim \frac{\sum_{i=1}^{n_A}(y_i - \widehat{\mu}_Y)^2}{\chi^2_{(n_A-1)}} \qquad = \chi^{-2}_{(n_A-1)} s_Y^2, \\ \mu_X|\sigma_X^2, x &\sim N_1(\widehat{\mu}_X, \sigma_X^2/n_B) \\ \sigma_X^2|x &\sim \frac{\sum_{i=1}^{n_B}(x_i - \widehat{\mu}_X)^2}{\chi^2_{(n_B-1)}} \qquad = \chi^{-2}_{(n_B-1)} s_X^2, \\ \rho &\sim f_\rho, \end{aligned}$$

with $\widehat{\mu}_Y = \frac{1}{n_A}\sum_{i=1}^{n_A} y_i$ and $\widehat{\mu}_X = \frac{1}{n_B}\sum_{i=1}^{n_B} x_i$.

Basic S-PLUS Routines

EVALprior

```
function(rho, n = 5000, k = 50, m = 5)
{
#NAME: EVALprior()
# Simulation study to evaluate the performance of statistical
# matching techniques when some prior information can be used
#################Preparing the data##################################
number <- length(rho)
mu <- c(0, 0, 0, 0)
sigmaZXY <- cbind(c(1, 0.2, 0.5, 0.8), c(0.2, 1, 0.5, 0.6),
    c(0.5, 0.5, 1, 0.8), c(0.8, 0.6, 0.8, 1))
print(sigmaZXY)
rNORM <- matrix(NA, (k * number), 27)
rNIBAS <- matrix(NA, (k * number), 27)
rMICE <- matrix(NA, (k * number), 27)
rRIEPS <- matrix(NA, (k * number), 27)
colname <- c("rhoXYgZ", "sigmaXY", rep("muX", m), rep("muY", m),
    rep("s2X", m), rep("s2Y", m), rep("r", m))
dimnames(rNORM) <- list(NULL, colname)
dimnames(rNIBAS) <- list(NULL, colname)
dimnames(rMICE) <- list(NULL, colname)
dimnames(rRIEPS) <- list(NULL, colname)
library(norm)
rngseed(12324545)
library(mice)
seed <- round(runif((k * number), 0, 1000))
now <- proc.time()
count <- 0
l <- 1
```

```
while(l <= number) {
    rhoXYgZ <- rho[l]
    j <- 1
    while(j <= k) {
        count <- count + 1
        set.seed(seed[count])
        ZXY <- rmvnorm(n, mu, sigmaZXY)
        ZXYNA <- ZXY
        ZXYNA[1:(n/2), 3] <- NA
        ZXYNA[(n/2 + 1):n, 4] <- NA
        A <- ZXYNA[1:(n/2),  ]
        B <- ZXYNA[(n/2 + 1):n,  ]
        SIGZXY <- (var(ZXY) * (n - 1))/n
        rNORM[count, 1] <- rhoXYgZ
        rNORM[count, 2] <- SIGZXY[3, 4]
        rNIBAS[count, 1] <- rhoXYgZ
        rNIBAS[count, 2] <- SIGZXY[3, 4]
        rMICE[count, 1] <- rhoXYgZ
        rMICE[count, 2] <- SIGZXY[3, 4]
        rRIEPS[count, 1] <- rhoXYgZ
        rRIEPS[count, 2] <- SIGZXY[3, 4]
#Calculation of the regression coefficients according to Rubin#####
        regressA <- lm(A[, 4] ~ A[, 1:2])
        resA <- residuals(regressA)
        sigmaYgZ <- sum(resA^2)/(n/2 - 3)
        betaYZ <- coef(regressA)
        regressB <- lm(B[, 3] ~ B[, 1:2])
        resB <- residuals(regressB)
        sigmaXgZ <- sum(resB^2)/(n/2 - 3)
        betaXZ <- coef(regressB)
        sigmaXYgZ <- rhoXYgZ * sqrt(sigmaYgZ * sigmaXgZ)
        betaXYgZ <- sigmaXYgZ/sigmaYgZ
        betaYXgZ <- sigmaXYgZ/sigmaXgZ
        betaA <- betaXZ - betaXYgZ * betaYZ
        betaB <- betaYZ - betaYXgZ * betaXZ
        betaA[4] <- betaXYgZ
        betaB[4] <- betaYXgZ
#Imputation via RIEPS##############################################
        XApred <- (cbind(rep(1, n/2), A[, 1:2], A[, 4])) %*% betaA
        YBpred <- (cbind(rep(1, n/2), B[, 1:3])) %*% betaB
        XBpred <- (cbind(rep(1, n/2), B[, 1:2], YBpred)) %*% betaA
        YApred <- (cbind(rep(1, n/2), A[, 1:2], XApred)) %*% betaB
        sigmaXgYZ <- sum((B[, 3] - XBpred)^2)/(n/2 - 4)
        sigmaYgXZ <- sum((A[, 4] - YApred)^2)/(n/2 - 4)
        for(i in 1:5) {
            A[, 3] <- XApred + rnorm((n/2), 0, sqrt(sigmaXgYZ))
            B[, 4] <- YBpred + rnorm((n/2), 0, sqrt(sigmaYgXZ))
            ZXYnew <- rbind(A, B)
            rRIEPS[count, (2 + i)] <- mean(ZXYnew[, 3])
```

```
            rRIEPS[count, (7 + i)] <- mean(ZXYnew[, 4])
            rRIEPS[count, (12 + i)] <- (var(ZXYnew[, 3])*(n - 1))/n
            rRIEPS[count, (17 + i)] <- (var(ZXYnew[, 4])*(n - 1))/n
            rRIEPS[count, (22 + i)] <- cor(ZXYnew)[3, 4]
        }
#Imputation via DA according to Schafer###########################
        varB <- (var(B[, 1:3]) * (n/2 - 1))/(n/2)
        varA <- (var(cbind(A[, 1:2], A[, 4])) * (n/2 - 1))/(n/2)
        SIGXYgZ <- cbind(c(sigmaXgZ, sigmaXYgZ), c(sigmaXYgZ,
            sigmaYgZ))
        sigma12 <- cbind(varB[1:2, 3], varA[1:2, 3])
        sigmaXY <- SIGXYgZ + t(sigma12) %*% solve(SIGZXY[1:2, 1:2])
            %*% sigma12
        sigmaXY <- sigmaXY[1, 2]
        prepare <- prelim.norm(ZXYNA)
        thetahat <- em.norm(prepare)
        thetalist <- getparam.norm(prepare, thetahat, corr = F)
        thetalist$sigma[4, 3] <- sigmaXY    #tweak the parameters
        thetalist$sigma[3, 4] <- sigmaXY
        thetahat <- mkparam.norm(prepare, thetalist)
        theta <- da.norm(prepare, thetahat, steps = 100)
            #yields a burn in period of 100
        for(i in 1:m) {
            ZXYnew <- imp.norm(prepare, theta, ZXYNA)
            theta <- da.norm(prepare, theta, steps = 50)
            rNORM[count, (2 + i)] <- mean(ZXYnew[, 3])
            rNORM[count, (7 + i)] <- mean(ZXYnew[, 4])
            rNORM[count, (12 + i)] <- (var(ZXYnew[, 3])*(n - 1))/n
            rNORM[count, (17 + i)] <- (var(ZXYnew[, 4])*(n - 1))/n
            rNORM[count, (22 + i)] <- cor(ZXYnew)[3, 4]
        }
#Imputation via the noniterative Bayesian model##################'
        for(i in 1:m) {
            sigmaYgZ <- sum(resA^2)/rchisq(1, (n/2 - 3))
            sigmaXgZ <- sum(resB^2)/rchisq(1, (n/2 - 3))
            sigmaXYgZ <- rhoXYgZ * sqrt(sigmaYgZ * sigmaXgZ)
            betaXYgZ <- sigmaXYgZ/sigmaYgZ
            betaYXgZ <- sigmaXYgZ/sigmaXgZ
            ZA <- cbind(c(rep(1, n/2)), A[, 1:2])
            ZB <- cbind(c(rep(1, n/2)), B[, 1:2])
            noteA <- solve(t(ZA) %*% ZA)
            noteB <- solve(t(ZB) %*% ZB)
            ZVbetaYZ <- t(rmvnorm(1, betaYZ, sigmaYgZ * noteA))
            ZVbetaXZ <- t(rmvnorm(1, betaXZ, sigmaXgZ * noteB))
            muAYgZ <- ZA %*% ZVbetaYZ
            muAXgZ <- ZA %*% ZVbetaXZ
            muBYgZ <- ZB %*% ZVbetaYZ
            muBXgZ <- ZB %*% ZVbetaXZ
            A[, 3] <- rnorm(n/2, (muAXgZ + betaXYgZ * (A[, 4] -
```

```
                muAYgZ)), sqrt(sigmaXgZ * (1 - rhoXYgZ^2)))
            B[, 4] <- rnorm(n/2, (muBYgZ + betaYXgZ * (B[, 3] -
                muBXgZ)), sqrt(sigmaYgZ * (1 - rhoXYgZ^2)))
            ZXYnew <- rbind(A, B)
            rNIBAS[count, (2 + i)] <- mean(ZXYnew[, 3])
            rNIBAS[count, (7 + i)] <- mean(ZXYnew[, 4])
            rNIBAS[count, (12 + i)] <- (var(ZXYnew[, 3])*(n - 1))/n
            rNIBAS[count, (17 + i)] <- (var(ZXYnew[, 4])*(n - 1))/n
            rNIBAS[count, (22 + i)] <- cor(ZXYnew)[3, 4]
        }
#Imputation via MICE##################################################
        MICEimp <- mice(ZXYNA)  #defaults are used
        for(i in 1:m) {
            ZXYnew <- complete(MICEimp, i)
            rMICE[count, (2 + i)] <- mean(ZXYnew[, 3])
            rMICE[count, (7 + i)] <- mean(ZXYnew[, 4])
            rMICE[count, (12 + i)] <- (var(ZXYnew[, 3])*(n - 1))/n
            rMICE[count, (17 + i)] <- (var(ZXYnew[, 4])*(n - 1))/n
            rMICE[count, (22 + i)] <- cor(ZXYnew)[3, 4]
        }
#do it again##########################################################
        j <- j + 1
    }
    l <- l + 1
}
speed <- proc.time() - now
print(speed)
list(rNORM, rNIBAS, rMICE, rRIEPS)
}
```

EVALcf

```
function(n = 5000, nhelp = 50, k = 50, m = 5)
{
# NAME: EVALcf()
# Simulation study to evaluate the performance of the matching
# techniques when a third data source can be used
####################### Preparing the data###########################
   mu <- c(0, 0, 0, 0)
   sigmaZXY <- cbind(c(1, 0.2, 0.5, 0.8), c(0.2, 1, 0.5, 0.6),
       c(0.5, 0.5, 1, 0.8), c(0.8, 0.6, 0.8, 1))
   print(sigmaZXY)
   rNORM <- matrix(NA, k, 7)
   rNIBAS <- matrix(NA, k, 7)
   rRIEPS <- matrix(NA, k, 7)
   rMICE <- matrix(NA, k, 7)
```

```
    dimnames(rNORM) <- list(NULL, c("rhoXYgZ", "sigmaXY", "r", "r",
        "r", "r", "r"))
    dimnames(rNIBAS) <- list(NULL, c("rhoXYgZ", "sigmaXY", "r", "r",
        "r", "r", "r"))
    dimnames(rRIEPS) <- list(NULL, c("rhoXYgZ", "sigmaXY", "r", "r",
        "r", "r", "r"))
    dimnames(rMICE) <- list(NULL, c("rhoXYgZ", "sigmaXY", "r", "r",
        "r", "r", "r"))
    library(norm)
    rngseed(12324545)
    library(mice)
    seed <- round(runif(k, 0, 1000))
    now <- proc.time()
    j <- 1
    count <- 1
    while(j <= k) {
        print(paste("Please wait -- the", j, "th run is on..."))
        set.seed(seed[j])
        ZXY <- rmvnorm((n + nhelp), mu, sigmaZXY)
        ZXYhelp <- ZXY[1:nhelp, ]
        ZXY <- ZXY[(nhelp + 1):(n + nhelp), ]
        ZXYNA <- ZXY
        ZXYNA[1:(n/2), 3] <- NA
        ZXYNA[(n/2 + 1):n, 4] <- NA
        ZXYNAorg <- rbind(ZXYhelp, ZXYNA)
        A <- ZXYNA[1:(n/2), ]
        B <- ZXYNA[(n/2 + 1):n, ]
        SIGZXY <- var(ZXY)
#Use control file for estimation of rhoXYgZ########################
        SIGZXYhelp <- var(ZXYhelp)
        sigmaXYgZ <- SIGZXYhelp[3:4, 3:4] - SIGZXYhelp[3:4, 1:2] %*%
            solve(SIGZXYhelp[1:2, 1:2]) %*% SIGZXYhelp[1:2, 3:4]
        rhoXYgZ <- sigmaXYgZ[1, 2]/sqrt(sigmaXYgZ[1, 1] *
            sigmaXYgZ[2, 2])
        print(paste("sigmaXY", round(SIGZXYhelp[3, 4], 4), "rhoXYgZ",
            round(rhoXYgZ, 4)))
        rNORM[count, 1] <- rhoXYgZ
        rNORM[count, 2] <- SIGZXYhelp[3, 4]
        rNIBAS[count, 1] <- rhoXYgZ
        rNIBAS[count, 2] <- SIGZXYhelp[3, 4]
        rMICE[count, 1] <- rhoXYgZ
        rMICE[count, 2] <- SIGZXYhelp[3, 4]
        rRIEPS[count, 1] <- rhoXYgZ
        rRIEPS[count, 2] <- SIGZXYhelp[3, 4]
#Calculate the regression coefficients according to Rubin#########
        regressA <- lm(A[, 4] ~ A[, 1:2])
        resA <- residuals(regressA)
        sigmaYgZ <- sum(resA^2)/(n/2 - 3)
        betaYZ <- coef(regressA)
```

```
        regressB <- lm(B[, 3] ~ B[, 1:2])
        resB <- residuals(regressB)
        sigmaXgZ <- sum(resB^2)/(n/2 - 3)
        betaXZ <- coef(regressB)
        sigmaXYgZ <- rhoXYgZ * sqrt(sigmaYgZ * sigmaXgZ)
        betaXYgZ <- sigmaXYgZ/sigmaYgZ
        betaYXgZ <- sigmaXYgZ/sigmaXgZ
        betaA <- betaXZ - betaXYgZ * betaYZ
        betaB <- betaYZ - betaYXgZ * betaXZ
        betaA[4] <- betaXYgZ
        betaB[4] <- betaYXgZ
#Imputation via RIEPS################################################
        XApred <- (cbind(rep(1, n/2), A[, 1:2], A[, 4])) %*% betaA
        YBpred <- (cbind(rep(1, n/2), B[, 1:3])) %*% betaB
        XBpred <- (cbind(rep(1, n/2), B[, 1:2], YBpred)) %*% betaA
        YApred <- (cbind(rep(1, n/2), A[, 1:2], XApred)) %*% betaB
        sigmaXgYZ <- sum((B[, 3] - XBpred)^2)/(n/2 - 4)
        sigmaYgXZ <- sum((A[, 4] - YApred)^2)/(n/2 - 4)
        for(i in 1:5) {
            A[, 3] <- XApred + rnorm((n/2), 0, sqrt(sigmaXgYZ))
            B[, 4] <- YBpred + rnorm((n/2), 0, sqrt(sigmaYgXZ))
            ZXYnew <- rbind(A, B)
            rRIEPS[count, (2 + i)] <- cor(ZXYnew)[3, 4]
        }
#Imputation via DA according to Schafer##############################
        prepare <- prelim.norm(ZXYNAorg)
        thetahat <- em.norm(prepare)
        theta <- da.norm(prepare, thetahat, steps = 100)
            #yields a burn in period of 100
        for(i in 1:m) {
            ZXYnew <- imp.norm(prepare, theta, ZXYNAorg)
            ZXYnew <- ZXYnew[(nhelp + 1):(n + nhelp), ]
            theta <- da.norm(prepare, theta, steps = 50)
            rNORM[count, (2 + i)] <- cor(ZXYnew)[3, 4]
        }
#Imputation via the noniterative Bayesian model#####################
        for(i in 1:m) {
            sigmaYgZ <- sum(resA^2)/rchisq(1, (n/2 - 3))
            sigmaXgZ <- sum(resB^2)/rchisq(1, (n/2 - 3))
            sigmaXYgZ <- rhoXYgZ * sqrt(sigmaYgZ * sigmaXgZ)
            betaXYgZ <- sigmaXYgZ/sigmaYgZ
            betaYXgZ <- sigmaXYgZ/sigmaXgZ
            ZA <- cbind(c(rep(1, n/2)), A[, 1:2])
            ZB <- cbind(c(rep(1, n/2)), B[, 1:2])
            noteA <- solve(t(ZA) %*% ZA)
            noteB <- solve(t(ZB) %*% ZB)
            ZVbetaYZ <- t(rmvnorm(1, betaYZ, sigmaYgZ * noteA))
            ZVbetaXZ <- t(rmvnorm(1, betaXZ, sigmaXgZ * noteB))
            muAYgZ <- ZA %*% ZVbetaYZ
```

```
            muAXgZ <- ZA %*% ZVbetaXZ
            muBYgZ <- ZB %*% ZVbetaYZ
            muBXgZ <- ZB %*% ZVbetaXZ
            A[, 3] <- rnorm(n/2, (muAXgZ + betaXYgZ * (A[, 4] -
                muAYgZ)), sqrt(sigmaXgZ * (1 - rhoXYgZ^2)))
            B[, 4] <- rnorm(n/2, (muBYgZ + betaYXgZ * (B[, 3] -
                muBXgZ)), sqrt(sigmaYgZ * (1 - rhoXYgZ^2)))
            ZXYnew <- rbind(A, B)
            rNIBAS[count, (2 + i)] <- cor(ZXYnew)[3, 4]
        }
#Imputation via MICE###################################################
        MICEimp <- mice(ZXYNAorg, maxit=150)
        for(i in 1:m) {
            ZXYnew <- complete(MICEimp, i)
            ZXYnew <- ZXYnew[(nhelp + 1):(n + nhelp),  ]
            rMICE[count, (2 + i)] <- cor(ZXYnew)[3, 4]
        }
#do it again###########################################################
        j <- j + 1
        count <- count + 1
    }
    speed <- proc.time() - now
    print(speed)
    list(rNORM, rNIBAS, rMICE, rRIEPS)
}
```

NIBAS

```
function(dataA, dataB, k, q, data.help = 0, nohelp = T)
{
# NAME:NIBAS()
# Calculation of the noniterative multivariate imputations NIBAS
# to perform the fusion of the media data with prior of zero or
# auxiliary information k common variables, q specific variables X
# and p specific variables Y
#######################################################################
   datanew <- cbind(dataA, dataB)
   nA <- length(dataA[, 1])
   nB <- length(dataB[, 1])
   r <- length(dataA[1,  ])
   p <- r - k - q
   dataimp <- list(datanew)
#calculate multiple regression########################################
   ZA <- cbind(c(rep(1, nA)), dataA[, 1:k])
   ZB <- cbind(c(rep(1, nB)), dataB[, 1:k])
   betaYZ <- solve(t(ZA) %*% ZA) %*% t(ZA) %*% dataA[,
```

```
            (k + q + 1):r]
        betaXZ <- solve(t(ZB) %*% ZB) %*% t(ZB) %*% dataB[,
            (k + 1):(k + q)]
        SY <- t(dataA[, (k + q + 1):r] - ZA %*% betaYZ) %*% (dataA[,
            (k + q + 1):r] - ZA %*% betaYZ)
        SX <- t(dataB[, (k + 1):(k + q)] - ZB %*% betaXZ) %*% (dataB[,
            (k + 1):(k + q)] - ZB %*% betaXZ)
    #prior of rhoXYgZ or use of auxiliary files #######################
        if(nohelp) {
            rhoXYgZ <- matrix(0, q, p)
        }
        else {
            help <- data.help
            ncomplete <- length(help[, 1])
            varhelp <- (var(help) * (ncomplete - 1))/ncomplete
            sigmaXYgZ.mini <- varhelp[(k + 1):r, (k + 1):r] -
                varhelp[(k + 1):r, 1:k] %*% solve(varhelp[1:
                k, 1:k]) %*% varhelp[1:k, (k + 1):r]
            rhoXYgZ <- bedkor(sigmaXYgZ.mini)
            rhoXYgZ <- rhoXYgZ[1:q, (q + 1):(q + p)]
        }
        vA <- nA - ((k + 1) + p) + 1
        vB <- nB - ((k + 1) + q) + 1
        sigmaYY <- solve(wishart(solve(SY), vA))
        sigmaXX <- solve(wishart(solve(SX), vB))
    #generate the random numbers for betaYZ and betaXY################
        varbetaYZ <- kronecker(sigmaYY, solve(t(ZA) %*% ZA))
        varbetaXZ <- kronecker(sigmaXX, solve(t(ZB) %*% ZB))
        ZVbetaYZ <- as.vector(rmvnorm(1, as.vector(betaYZ), varbetaYZ))
        ZVbetaYZ <- matrix(ZVbetaYZ, nrow = (k + 1))
        ZVbetaXZ <- as.vector(rmvnorm(1, as.vector(betaXZ), varbetaXZ))
        ZVbetaXZ <- matrix(ZVbetaXZ, nrow = (k + 1))
    #calculate the matrix sigmaXYgZ####################################
        sigma12 <- rhoXYgZ * sqrt(diag(sigmaXX) %*% t(diag(sigmaYY)))
    # generate the random numbers for XgY and YgX ####################
        muXgY <- ZA %*% ZVbetaXZ + (dataA[, (k + q + 1):r] - ZA %*%
            ZVbetaYZ) %*% solve(sigmaYY) %*% t(sigma12)
        varXgY <- sigmaXX - sigma12 %*% solve(sigmaYY) %*% t(sigma12)
        muYgX <- ZB %*% ZVbetaYZ + (dataB[, (k + 1):(k + q)] - ZB %*%
            ZVbetaXZ) %*% solve(sigmaXX) %*% sigma12
        varYgX <- sigmaYY - t(sigma12) %*% solve(sigmaXX) %*% sigma12
        dataA[, (k + 1):(k + q)] <- rmvnorm(nA, muXgY, varXgY)
        dataB[, (k + q + 1):r] <- rmvnorm(nB, muYgX, varYgX)
        dataimp <- rbind(dataA, dataB)
        dataimp
}
```

Tables

Table A.1: RIEPS: Regression imputation with stochastic residual

θ	$\rho^{prio}_{XY\|Z}$	$\widehat{E}(\widehat{\theta}^{MI})$	$\widehat{Var}(\widehat{\theta}^{MI})$	$\widehat{E}(W)$	$\widehat{E}(B)$	Cvg.
μ_X	-0.9	-0.000072	0.000288	0.000191	0.000002	43
μ_Y	-0.9	-0.002795	0.000193	0.000197	0.000000	46
σ_{XX}	-0.9	0.953235	0.000659	0.000364	0.000008	20
σ_{YY}	-0.9	0.984562	0.000576	0.000388	0.000002	40
ρ_{XY}	-0.9	0.318571	0.000329	0.000200	0.000003	38
μ_X	-0.8	0.004946	0.000271	0.000187	0.000007	44
μ_Y	-0.8	0.001644	0.000219	0.000196	0.000002	48
σ_{XX}	-0.8	0.933496	0.000952	0.000349	0.000027	10
σ_{YY}	-0.8	0.981274	0.000658	0.000385	0.000007	37
ρ_{XY}	-0.8	0.354740	0.000321	0.000200	0.000012	36
μ_X	-0.4	0.000155	0.000364	0.000190	0.000040	44
μ_Y	-0.4	0.000275	0.000236	0.000198	0.000010	48
σ_{XX}	-0.4	0.951647	0.000755	0.000363	0.000118	25
σ_{YY}	-0.4	0.988466	0.000543	0.000391	0.000041	44
ρ_{XY}	-0.4	0.472043	0.000181	0.000200	0.000060	42
μ_X	0.0	-0.003450	0.000321	0.000200	0.000061	48
μ_Y	0.0	-0.001330	0.000304	0.000199	0.000014	44
σ_{XX}	0.0	1.000759	0.001033	0.000401	0.000195	45
σ_{YY}	0.0	0.996230	0.000455	0.000397	0.000063	49
ρ_{XY}	0.0	0.583517	0.000112	0.000200	0.000081	49
μ_X	0.4	0.004937	0.000402	0.000192	0.000038	44
μ_Y	0.4	0.001608	0.000240	0.000197	0.000012	46
σ_{XX}	0.4	0.961703	0.000724	0.000370	0.000145	28
σ_{YY}	0.4	0.986980	0.000593	0.000390	0.000047	41
ρ_{XY}	0.4	0.721828	0.000092	0.000200	0.000109	26
μ_X	0.8	0.001054	0.000361	0.000186	0.000006	41
μ_Y	0.8	-0.001509	0.000240	0.000196	0.000002	48
σ_{XX}	0.8	0.929749	0.000697	0.000346	0.000022	5
σ_{YY}	0.8	0.980450	0.000486	0.000385	0.000007	39
ρ_{XY}	0.8	0.863747	0.000058	0.000200	0.000031	0
μ_X	0.9	0.003376	0.000327	0.000190	0.000002	45
μ_Y	0.9	-0.000192	0.000206	0.000197	0.000001	49
σ_{XX}	0.9	0.951874	0.000489	0.000363	0.000007	17
σ_{YY}	0.9	0.985290	0.000294	0.000388	0.000002	44
ρ_{XY}	0.9	0.883325	0.000065	0.000200	0.000010	0

Table A.2: NORM: Data augmentation imputation

θ	$\rho^{prio}_{XY\mid Z}$	$\widehat{E}(\widehat{\theta}^{MI})$	$\widehat{Var}(\widehat{\theta}^{MI})$	$\widehat{E}(W)$	$\widehat{E}(B)$	Cvg.
μ_X	-0.9	-0.005575	0.000402	0.000200	0.000117	47
μ_Y	-0.9	0.001755	0.000273	0.000200	0.000031	47
σ_{XX}	-0.9	0.998612	0.000644	0.000399	0.000333	49
σ_{YY}	-0.9	0.998306	0.000709	0.000399	0.000115	46
ρ_{XY}	-0.9	0.300662	0.000255	0.000200	0.000169	48
μ_X	-0.8	0.004572	0.000285	0.000201	0.000113	48
μ_Y	-0.8	-0.001671	0.000178	0.000200	0.000029	50
σ_{XX}	-0.8	1.002647	0.000784	0.000403	0.000354	48
σ_{YY}	-0.8	1.001178	0.000622	0.000401	0.000116	47
ρ_{XY}	-0.8	0.333906	0.000578	0.000200	0.000287	45
μ_X	-0.4	0.002155	0.000285	0.000200	0.000128	48
μ_Y	-0.4	-0.001346	0.000129	0.000200	0.000032	50
σ_{XX}	-0.4	0.997928	0.000815	0.000399	0.000286	48
σ_{YY}	-0.4	1.000767	0.000646	0.000401	0.000116	47
ρ_{XY}	-0.4	0.458843	0.002748	0.000200	0.001721	40
μ_X	0.0	0.000726	0.000457	0.000200	0.000110	46
μ_Y	0.0	-0.002215	0.000199	0.000199	0.000036	48
σ_{XX}	0.0	0.998969	0.000754	0.000400	0.000331	48
σ_{YY}	0.0	0.997430	0.000454	0.000398	0.000120	47
ρ_{XY}	0.0	0.592536	0.006090	0.000200	0.003584	42
μ_X	0.4	0.002552	0.000421	0.000201	0.000112	47
μ_Y	0.4	0.002418	0.000182	0.000201	0.000038	48
σ_{XX}	0.4	1.003126	0.000700	0.000403	0.000351	47
σ_{YY}	0.4	1.003375	0.000549	0.000403	0.000123	49
ρ_{XY}	0.4	0.695367	0.003710	0.000200	0.004196	43
μ_X	0.8	-0.004204	0.000223	0.000200	0.000113	48
μ_Y	0.8	-0.001749	0.000240	0.000199	0.000031	47
σ_{XX}	0.8	0.998251	0.000765	0.000399	0.000356	50
σ_{YY}	0.8	0.996319	0.000435	0.000397	0.000107	49
ρ_{XY}	0.8	0.832996	0.000528	0.000200	0.001642	40
μ_X	0.9	0.001665	0.000295	0.000202	0.000109	49
μ_Y	0.9	-0.001133	0.000231	0.000200	0.000030	48
σ_{XX}	0.9	1.008262	0.000733	0.000407	0.000367	48
σ_{YY}	0.9	1.001698	0.000437	0.000402	0.000109	47
ρ_{XY}	0.9	0.856310	0.000187	0.000200	0.000942	44

Table A.3: NIBAS: Noniterative multivariate imputation

θ	$\rho^{prio}_{XY\|Z}$	$\widehat{E}(\widehat{\theta}^{MI})$	$\widehat{Var}(\widehat{\theta}^{MI})$	$\widehat{E}(W)$	$\widehat{E}(B)$	Cvg.
μ_X	-0.9	-0.005892	0.000401	0.000200	0.000137	47
μ_Y	-0.9	0.001723	0.000265	0.000199	0.000031	47
σ_{XX}	-0.9	0.997855	0.000637	0.000399	0.000354	48
σ_{YY}	-0.9	0.997091	0.000719	0.000398	0.000121	46
ρ_{XY}	-0.9	0.303395	0.000193	0.000200	0.000125	49
μ_X	-0.8	0.004032	0.000305	0.000201	0.000123	49
μ_Y	-0.8	-0.002167	0.000183	0.000200	0.000031	49
σ_{XX}	-0.8	1.006236	0.000818	0.000405	0.000360	47
σ_{YY}	-0.8	1.002398	0.000630	0.000402	0.000106	46
ρ_{XY}	-0.8	0.340460	0.000234	0.000200	0.000114	47
μ_X	-0.4	0.002812	0.000311	0.000200	0.000136	48
μ_Y	-0.4	-0.001652	0.000128	0.000200	0.000029	50
σ_{XX}	-0.4	0.997790	0.000873	0.000399	0.000292	46
σ_{YY}	-0.4	1.000711	0.000654	0.000401	0.000117	47
ρ_{XY}	-0.4	0.461598	0.000204	0.000200	0.000162	49
μ_X	0.0	0.000956	0.000429	0.000199	0.000127	47
μ_Y	0.0	-0.002510	0.000185	0.000200	0.000030	47
σ_{XX}	0.0	0.997318	0.000766	0.000398	0.000355	46
σ_{YY}	0.0	0.997999	0.000463	0.000399	0.000124	49
ρ_{XY}	0.0	0.582467	0.000077	0.000200	0.000157	50
μ_X	0.4	0.001719	0.000396	0.000201	0.000137	46
μ_Y	0.4	0.002670	0.000191	0.000200	0.000034	49
σ_{XX}	0.4	1.004629	0.000692	0.000404	0.000302	48
σ_{YY}	0.4	1.001707	0.000637	0.000402	0.000116	47
ρ_{XY}	0.4	0.704841	0.000096	0.000200	0.000208	49
μ_X	0.8	-0.003668	0.000209	0.000200	0.000124	49
μ_Y	0.8	-0.001703	0.000224	0.000199	0.000034	49
σ_{XX}	0.8	0.998721	0.000755	0.000399	0.000292	47
σ_{YY}	0.8	0.996787	0.000448	0.000398	0.000128	50
ρ_{XY}	0.8	0.825035	0.000101	0.000200	0.000375	46
μ_X	0.9	0.002187	0.000303	0.000202	0.000124	46
μ_Y	0.9	-0.001488	0.000214	0.000200	0.000037	49
σ_{XX}	0.9	1.009214	0.000682	0.000408	0.000329	47
σ_{YY}	0.9	1.000408	0.000430	0.000401	0.000123	48
ρ_{XY}	0.9	0.855077	0.000080	0.000200	0.000692	45

Table A.4: MICE: Multiple imputations using chained equations

θ	$\rho^{prio}_{XY\|Z}$	$\widehat{E}(\widehat{\theta}^{MI})$	$\widehat{Var}(\widehat{\theta}^{MI})$	$\widehat{E}(W)$	$\widehat{E}(B)$	Cvg.
μ_X	-0.9	0.004286	0.000340	0.000199	0.000078	46
μ_Y	-0.9	0.002169	0.000174	0.000200	0.000023	50
σ_{XX}	-0.9	0.996966	0.000683	0.000398	0.000232	47
σ_{YY}	-0.9	0.997882	0.000403	0.000398	0.000083	49
ρ_{XY}	-0.9	0.580046	0.000159	0.000200	0.000661	0
μ_X	0.0	0.004522	0.000433	0.000200	0.000089	46
μ_Y	0.0	0.001945	0.000258	0.000199	0.000025	46
σ_{XX}	0.0	1.001917	0.000800	0.000402	0.000289	47
σ_{YY}	0.0	0.992642	0.000561	0.000394	0.000069	46
ρ_{XY}	0.0	0.575082	0.000214	0.000200	0.000700	48
μ_X	0.9	-0.002389	0.000282	0.000200	0.000067	49
μ_Y	0.9	-0.000093	0.000272	0.000200	0.000028	48
σ_{XX}	0.9	0.999727	0.000842	0.000400	0.000242	48
σ_{YY}	0.9	1.000336	0.000660	0.000401	0.000082	44
ρ_{XY}	0.9	0.577289	0.000256	0.000200	0.000830	0

References

Adamek, J.C. (1994) Fusion: Combining Data from Separate Sources, *Marketing Research: A Magazine of Management and Applications*, **6**, 48–50.

Alter, H. (1974) Creation of a Synthetic Data Set by Linking Records of the Canadian Survey of Consumer Finances with the Family Expenditure Survey 1970, *Annals of Economic and Social Measurement*, **3/2**, 373–394.

Aluja-Banet, T., Thio, S. (2001) Survey Data Fusion, *Bulletin of Sociological Methodology*, **72**, 20–36.

Anderson, T.W. (1984) *An Introduction to Multivariate Statistical Analysis.* Wiley, New York.

Antoine, J. (1987) A Case Study Illustrating the Objectives and Perspectives of Fusion Techniques, *Readership Research: Theory and Practice* (ed. Henry, H.), 336–351, Elsevier Science, Amsterdam.

Antoine, J., Santini, G. (1987) Fusion Techniques: Alternative to Single-Source Methods, *European Research*, **15**, 178–187.

Arbeitsgemeinschaft Media-Analyse e.V. (1996) *Regeln für die Handhabung von Zusammenführungen von MA-Daten mit Daten anderer Herkunft.*

Bacher, J. (2000) A Probabilistic Clustering Model for Variables of Mixed Type, *Quality and Quantity*, **34**, 223–235.

Baker, K. (1990) *The BARB/TGI Fusion. Technical Report on the Fusion Conducted in February/March 1990.* Ken Baker Associates, Ickenham, UK.

Baker, K., Harris, P., O'Brien, J. (1989) Data Fusion: An Appraisal and Experimental Evaluation, *Journal of the Market Research Society*, **31**, 153–212.

Bennike, S. (1987) Fusion – An Overview by an Outside Observer, *Readership Research: Theory and Practice* (ed. Henry, H.), 334-335, Elsevier Science, Amsterdam.

Berger, J.O. (1985) *Statistical Decision Theory and Bayesian Analysis.* Springer, New York.

Bergonier, H. (1968) Typological Analyses, *Seminar on Operational Research in Marketing*, ESOMAR, Neu Isenburg, 167–178.

Bergonier, H., Boucharenc, L., Irrmann, P. (1967) Une Nouvelle Méthode d'Analyse Globale des Résultats d'une Enquête Etablissement de Typologies, *Revue Francaise de Marketing*, **25**, 31–41.

Box, G.E.P., Tiao, G.C. (1992) *Bayesian Inference in Statistical Analysis.* Wiley, New York.

Brand, J.P.L. (1999) Development, Implementation and Evaluation of Multiple Imputation Strategies for the Statistical Analysis of Incomplete Data Sets, *Thesis Erasmus University Rotterdam*, Print Partners Ispkamp, Enschede, The Netherlands.

Buck, S. (1989) Single Source Data – the Theory and the Practice, *Journal of the Market Research Society*, **31**, 489–500.

Budd, E.C. (1971) The Creation of a Microdata File for Estimating the Size Distribution of Income, *Review of Income and Wealth*, **17**, 317–334.

Büning, H., Trenkler, G. (1978) *Nichtparametrische statistische Methoden.* De Gruyter, Berlin.

Carlin, B.P., Louis, T.A. (2000) *Bayes and Empirical Bayes Methods for Data Analysis.* Chapman and Hall, London.

Casella, G., George, E.I. (1992) Explaining the Gibbs Sampler, *The American Statistician*, **46**, 167–174.

Casella, G., Berger, R.L. (1990) *Statistical Inference.* Duxbury, Belmont, California.

Chen, M.H., Shao, Q.M., Ibrahim, J.G. (2000) *Monte Carlo Methods in Bayesian Computation.* Springer, New York.

Cox, D.R., Wermuth, N. (1996) *Multivariate Dependencies.* Chapman and Hall, London.

Darkow, M. (1996) Compatible or Not? Results of a Single Source Field Experiment Within a TV Audience Research Panel, *Marketing and Research Today*, August 1996, 150–161.

Dempster, A.P., Laird, N.M., Rubin, D.B. (1977), Maximum Likelihood Estimation from Incomplete Data via the EM Algorithm (with discussion), *Journal of the Royal Statistical Society B*, **39**, 1–38.

Everitt, B.S. (1984) *An Introduction to Latent Variable Models.* Chapman and Hall, London.

Fahrmeir, L., Hamerle, A., Tutz, G. (1996) *Multivariate Statistische Verfahren.* De Gruyter, Berlin.

Fellegi, I.P. (1977) Discussion, *Proceedings of the American Statistical Association, Social Statistics Section*, 762–764.

Gabler, S. (1997) Datenfusion, *ZUMA-Nachrichten*, **40**, 81–92.

Gelman, A., Carlin, J.B., Stern, H.S., Rubin, D.B. (1995) *Bayesian Data Analysis.* Chapman and Hall, London.

Gelman, A., King, G., Liu, C. (1998) Not Asked and Not Answered: Multiple Imputation for Multiple Surveys (with discussion), *Journal of the American Statistical Association*, **93**, 846–869.

Gilks, W.R., Richardson, S., Spiegelhalter, D.J. (1996) *Markov Chain Monte Carlo in Practice.* Chapman and Hall, London.

Goel, P.K., Ramalingam, T. (1989) *The Matching Methodology: Some Statistical Properties.* Lecture Notes in Statistics. Springer, Berlin.

Greene, W.H. (1997) *Econometric Analysis.* Prentice-Hall, Englewood Cliffs, New Jersey.

Groves, R.M., Dillman, D.A., Eltinge, J.L., Little, R.J.A. (2002) *Survey Nonresponse*, Wiley, New York.

Holland, P.W. (1986) Statistics and Causal Inference, *Journal of the American Statistical Association*, **81**, 945–960.

Horowitz, J.L., Manski, C.F. (2000a) Nonparametric Analysis of Randomized Experiments with Missing Covariate and Outcome Data, *Journal of the American Statistical Association*, **95**, 77–84.

Horowitz, J.L., Manski, C.F. (2000b) Rejoinder, *Journal of the American Statistical Association*, **95**, 87–88.

Judkins, D.R. (1998) Comment, *Journal of the American Statistical Association*, **93**, 861–864.

Kadane, J.B. (2001) Some Statistical Problems in Merging Data Files, *Journal of Official Statistics*, **17**, 423–433.

Kamakura, W.A., Wedel, M. (1997) Statistical Data Fusion for Cross-Tabulation, *Journal of Marketing Research*, **34**, 485–498.

Kennickell, A.B. (1991) Imputation of the 1989 Survey of Consumer Finances: Stochastic Relaxation and Multiple Imputation, *Proceedings of the Survey Research Methods Section, American Statistical Association*, 1–10.

Kennickell, A.B., McManus, D.A. (1994) Multiple Imputation of the 1983 and 1989 Waves of the SCF, *Proceedings of the Survey Research Methods Section, American Statistical Association*, 523–528.

Koschnick, W.J. (1995) *Standard-Lexikon für Mediaplanung und Mediaforschung in Deutschland.* K. G. Saur Verlag KG, München.

Kovacevic, M.S., Liu, T. (1994) Statistical Matching of Survey Datafiles: A Simulation Study, *Proceedings of the Section on Survey Research Methods, American Statistical Association*, 497–484.

Lee, H., Rancourt, E., Särndal C.E. (2002) Variance Estimation from Survey Data under Single Imputation, *Survey Nonresponse* (eds. Groves, R.M., Dillman, D.A., Eltinge, J.L., Little, R.J.A.), 315–328, Wiley, New York.

Li, K.H., Raghunathan, T.E., Rubin, D.B. (1991) Large-Sample Significance Levels from Multiply Imputed Data Using Moment-Based Statistics and an F Reference Distribution, *Journal of the American Statistical Association*, **86**, 1065–1073.

Little, R.J.A. (1986) Survey Nonresponse Adjustments, *International Statistical Review*, **54**, 139–157,

Little, R.J.A. (1988) Missing-Data Adjustments in Large Surveys, *Journal of Business and Economic Statistics*, **6**, 287–296.

Little, R.J.A., Rubin, D.B. (1987) *Statistical Analysis with Missing Data.* Wiley, New York.

Liu, T.P. (1998) A Categorical Constraints Guided Matching Algorithm, *Proceedings of the Survey Methods Section, Statistical Society of Canada*, 249–256.

Liu, T.P., Kovacevic, M.S. (1996) Categorically Constrained Matching, *Proceedings of the Survey Methods Section, Statistical Society of Canada*, 123–133.

Liu, T.P., Kovacevic, M.S. (1997) An Empirical Study on Categorically Constrained Matching, *Proceedings of the Survey Methods Section, Statistical Society of Canada*, 167–178.

Liu, T.P., Kovacevic, M.S. (1998) Categorical Matching and Constrained Rematching of Survey Datafiles, *Working Paper No. HSMD-98-008E*, Statistics Canada.

Manski, C.F. (1995) *Identification Problems in the Social Sciences.* Harvard University Press, Cambridge, Massachusetts.

Marker, D.A., Judkins, D.R., Winglee, M. (2002) Large-Scale Imputation for Complex Surveys, *Survey Nonresponse* (eds. Groves, R.M., Dillman, D.A., Eltinge, J.L., Little, R.J.A.), 329–341, Wiley, New York.

McLachlan, G.J., Krishnan, T. (1997) *The EM Algorithm and Extensions.* Wiley, New York.

Meng, X.L. (2002) A Congenial Overview and Investigation of Multiple Imputation Inferences under Uncongeniality, *Survey Nonresponse* (eds. Groves, R.M., Dillman, D.A., Eltinge, J.L., Little, R.J.A.), 343–356, Wiley, New York.

Meng, X.L., Rubin, D.B. (1992) Performing Likelihood Ratio Tests with Multiply-Imputed Data Sets, *Biometrica*, **79**, 103–111.

Mood, A.M., Graybill, F.A., Boes, D.C. (1974) *Introduction to the Theory of Statistics.* McGraw-Hill, New York.

Moriarity, C., Scheuren, F. (2001) Statistical Matching: A Paradigm for Assessing the Uncertainty in the Procedure, *Journal of Official Statistics*, **17**, 407–422.

Nordholt, E.S. (1998) Imputation: Methods, Simulation Experiments and Practical Examples, *International Statistical Review*, **66**, 157–180.

O'Brien, S. (1991) The Role of Data Fusion in Actionable Media Targeting in the 1990's, *Marketing and Research Today*, February 1991, 15–22.

Okner, B.A. (1972a) Constructing a New Data Base from Merging Microdata Sets: The 1966 Merge File, *Annals of Economic and Social Measurement*, **1**, 325–341.

Okner, B.A. (1972b) Reply and Comments, *Annals of Economic and Social Measurement*, **1**, 359–362.

Okner, B.A. (1974) Data Matching and Merging: An Overview, *Annals of Economic and Social Measurement*, **3**, 347–352.

Oudshoorn, K., van Buuren, S., van Rijckevorsel, J. (1999) Flexible Multiple Imputations by Chained Equations of the AVO-95 Survey, *TNO Report PG/VGZ/99.045*, Leiden.

Paass, G. (1986) Statistical Match: Evaluation of Existing Procedures and Improvements by Using Additional Information, *Micro-Analytic Sim-*

ulation Models to Support Social and Financial Policy (eds. Orcutt, Merz, and Quinke), Elsevier Science, Amsterdam.

Rässler, S. (2000) Ergänzung fehlender Daten in Umfragen, *Jahrbücher für Nationalökonomie und Statistik*, **220/1**, 64–94.

Rässler, S., Fleischer, K. (1998) Aspects Concerning Data Fusion Techniques, *ZUMA Nachrichten Spezial*, **4**, 317–333.

Rässler, S., Fleischer, K. (2000) An Evaluation of Data Fusion Techniques, *Proceedings of Statistics Canada Symposium 99*, 129–136.

Raetzel, D.F. (2000) Werbewirkungsanalyse mit fusionierten Daten. *Masters Thesis*, Marburg, Germany.

Raghunathan, T.E. (2000) Comment, *Journal of the American Statistical Association*, **95**, 85–87.

Raghunathan, T.E., Grizzle, J.E. (1995) A Split Questionnaire Survey Design, *Journal of the American Statistical Association*, **90**, 54–63.

Robert, C.P., Casella, G. (1999) *Monte Carlo Statistical Methods.* Springer, New York.

Roberts, A. (1994) Media Exposure and Consumer Purchasing: An Improved Data Fusion Technique, *Marketing and Research Today*, August 1994, 150–172.

Rodgers, W.L. (1984) An Evaluation of Statistical Matching, *Journal of Business and Econometric Statistics*, **2**, 91–102.

Rosenbaum, P.R. (1999) Choice as an Alternative to Control in Observational Studies, *Statistical Sciences*, **14**, 259–304.

Rosenbaum, P.R., Rubin, D.B. (1983) The Central Role of the Propensity Score in Observational Studies for Causal Effects, *Biometrika*, **70**, 41–55.

Rosenbaum, P.R., Rubin, D.B. (1985) Constructing a Control Group Using Multivariate Matched Sampling Methods that Incorporate the Propensity Score, *The American Statistician*, **39**, 33–38.

Rubin, D.B. (1974) Characterizing the Estimation of Parameters in Incomplete-Data Problems, *Journal of the American Statistical Association*, **69**, 467–474.

Rubin, D.B. (1986) Statistical Matching Using File Concatenation with Adjusted Weights and Multiple Imputations, *Journal of Business and Economic Statistics*, **4**, 87–95.

Rubin, D.B. (1987) *Multiple Imputation for Nonresponse in Surveys.* Wiley, New York.

Rubin, D.B. (1996) Multiple Imputation After 18+ Years (with discussion), *Journal of the American Statistical Association*, **91**, 473–489.

Rubin, D.B., Stern H.S., Vehovar, V. (1995) Handling "Don't Know" Slovenian Responses: The Case of the Slovenian Plebiscite, *Journal of the American Statistical Association*, **90**, 822–828.

Rubin, D.B., Thomas, N. (2000) Combining Propensity Score Matching with Additional Adjustment for Prognostic Covariates, *Journal of the American Statistical Association*, **95**, 573–585.

Ruggles, N.N., Ruggles, R. (1974) A Strategy for Merging and Matching Microdata Sets, *Annals of Economic and Social Measurement*, **3**, 353–371.

Ruggles, N.N., Ruggles, R., Wolf, E.D. (1977) Merging Microdata: Rationale, Practice and Testing, *Annals of Economic and Social Measurement*, **6**, 407–428.

Schafer, J.L. (1996) NORM: Multiple Imputation for Continuous Data Based on a Normal Model, Software Library written for S-PLUS, written in S-PLUS and FORTRAN-77, available at http://www.stat.psu.edu/~jls/.

Schafer, J.L. (1997) *Analysis of Incomplete Multivariate Data.* Chapman and Hall, London.

Schafer, J.L. (1999a) Multiple Imputation: A Primer, *Statistical Methods in Medical Research*, **8**, 3–15.

Schafer, J.L. (1999b) Multiple Imputation Under a Normal Model, Version 2, Software for Windows 95/98/NT, available at http://www.stat.psu.edu/~jls/misoftwa.html.

Schafer, J.L., Olsen, M.K. (1999) Modeling and Imputation of Semicontinuous Survey Variables, *Technical Report No. 00-39*, The Pennsylvania State University.

Shao, J. (2002) Replication Methods for Variance Estimation in Complex Surveys with Imputed Data, *Survey Nonresponse* (eds. Groves, R.M., Dillman, D.A., Eltinge, J.L., Little, R.J.A.), 303–314, Wiley, New York.

Scheler, H.-E. (1993) Der Datentransfer im Partnerschaftsmodell der Arbeitsgemeinschaft Media-Analyse, *Arbeitsgemeinschaft Media-Analyse e.V.*, Media-Micro-Census GmbH, Frankfurt.

Scheler, H.-E., Wiegand, J. (1987) A Report on Experiments in Fusion in the "Official" German Media Research (AG.MA), *Readership Research: Theory and Practice* (ed. Henry, H.), 352–360, Elsevier Science, Amsterdam.

Schenker, N., Welsh, A.H. (1988) Asymptotic Results for Multiple Imputation, *Annals of Statistics*, **16**, 1550–1566.

Sims, C.A. (1972a) Comments, *Annals of Economic and Social Measurement*, **1**, 343–345.

Sims, C.A. (1972b) Rejoinder, *Annals of Economic and Social Measurement*, **1**, 355–357.

Singh, A.C., Armstrong, J., Lemaître, G.E. (1988) Statistical Matching Using Log-Linear Imputation, *Proceedings of the Section on Survey Research Methods, American Statistical Association*, 672–677.

Singh, A.C., Mantel, H.J., Kinack, M.D., Rowe, G. (1993) Statistical Matching: Use of Auxiliary Information as an Alternative to the Conditional Independence Assumption, *Survey Methodology*, **19**, 59–79.

Song, J., Belin, T.R. (1999) Analysis of Incomplete High-Dimensional Multivariate Normal Data Using a Common Factor Model, *Proceedings of the American Statistical Association Section on Survey Research Methods*, 103–111.

Stuart, A., Ord, J.K. (1991) *Kendall's Advanced Theory of Statistics Vol. 2*. Edward Arnold, London.

Tanner, M.A. (1996) *Tools for Statistical Inference: Methods for the Exploration of Posterior Distributions and Likelihood Functions*. Springer, New York.

Tanner, M.A., Wong, W.H. (1987) The Calculation of Posterior Distributions by Data Augmentation (with discussion), *Journal of the American Statistical Association*, **82**, 528–550.

Tiao, G.C., Zellner, A. (1964) On the Bayesian Estimation of Multivariate Regression, *Journal of the Royal Statistical Society B*, **26**, 277–285.

Van Buuren, S., Oudshoorn, C.G.M. (2000) Multivariate Imputation by Chained Equations, *TNO Report PG/VGZ/00.038*, Leiden.

Van Buuren, S., Oudshoorn, K. (1999) Flexible Multivariate Imputation by MICE, *TNO Report PG/VGZ/99.054*, Leiden.

Van Dyk, D.A., Meng, X.L. (2001) The Art of Data Augmentation (with discussion), *Journal of Computational and Graphical Statistics*, **10**, 1–111.

Wendt, F. (1975) *Typologische Analyse Systeme in der deutschen Marktforschung*. Gruner und Jahr, Hamburg.

Wendt, F. (1976) *Beschreibung einer Fusion*. Gruner und Jahr, Schriftenreihe 21, Hamburg.

Wendt, F. (1979) *Bericht über ein Fusionsexperiment mit Daten der Niederländischen Mediauntersuchung NOP 77.* Bureau Wendt, Hamburg-Puidoux.

Wendt, F. (1980) *Zusammenführung von Daten – Voraussetzungen und Grenzen.* Gruner und Jahr, Schriftenreihe 28, Hamburg.

Wendt, F. (1986) Einige Gedanken zur Fusion, *Auf dem Wege zum Partnerschaftsmodell*, Arbeitsgemeinschaft Media-Analyse e.V., Media-Micro-Census GmbH, Frankfurt, 109–140.

Wendt, F. (2000) Bericht über die Durchführung der Fusionen zur MA 2000, *MA 2000*, Arbeitsgemeinschaft Media-Analyse e.V., Media-Micro-Census GmbH, Frankfurt, 1–61.

Wendt, F., Wendt-Maeder, I. (1979) *Vom Leser pro Nummer zur Nutzungswahrscheinlichkeit, Teil 1.* Arbeitsgemeinschaft Media-Analyse e.V. Schriften Band 5, Media-Micro-Census GmbH, Frankfurt.

Wendt, F., Wendt-Maeder, I. (1983) *Vom Leser pro Nummer zur Nutzungswahrscheinlichkeit, Teil 2.* Arbeitsgemeinschaft Media-Analyse e.V. Schriften Band 5, Media-Micro-Census GmbH, Frankfurt.

Wendt, F., Wendt-Maeder, I. (1986) *Vom Leser pro Nummer zur Nutzungswahrscheinlichkeit, Teil 3.* Arbeitsgemeinschaft Media-Analyse e.V. Schriften Band 5, Media-Micro-Census GmbH, Frankfurt.

Wendt-Maeder, I. (1976) *Fusionen im AG.MA-Bereich Bedingungen und Möglichkeiten.* Arbeitsgemeinschaft Media-Analyse e.V., Media-Micro-Census GmbH, Frankfurt.

Whittaker, J. (1990) *Graphical Models in Applied Multivariate Statistics.* Wiley, New York.

Wiegand, J. (1986) Combining Different Media Surveys, *Journal of the Market Research Society*, **28**, 189–208.

Wildner, R. (1994) Die Grenzen von Single Source, *planung and analyse*, **1**, 22–27.

Wildner, R. (2000) Measuring Advertising Effectiveness in a Sample Matched from a Consumer Panel and a Television Panel, *Proceedings of the Fifth International Conference on Logic and Methodology*, Cologne.

Woodbury, M.A. (1983) Statistical Record Matching for Files, *Incomplete Data in Sample Surveys 3, Proceedings of the Symposium* (eds. Madow, W.G., Olkin, I.), 173–181, Academic, New York.

Index

Lecture Notes in Statistics

For information about Volumes 1 to 115, please contact Springer-Verlag

116: Genshiro Kitagawa and Will Gersch, Smoothness Priors Analysis of Time Series. x, 261 pp., 1996.

117: Paul Glasserman, Karl Sigman, and David D. Yao (Editors), Stochastic Networks. xii, 298 pp., 1996.

118: Radford M. Neal, Bayesian Learning for Neural Networks. xv, 183 pp., 1996.

119: Masanao Aoki and Arthur M. Havenner, Applications of Computer Aided Time Series Modeling. ix, 329 pp., 1997.

120: Maia Berkane, Latent Variable Modeling and Applications to Causality. vi, 288 pp., 1997.

121: Constantine Gatsonis, James S. Hodges, Robert E. Kass, Robert McCulloch, Peter Rossi, and Nozer D. Singpurwalla (Editors), Case Studies in Bayesian Statistics, Volume III. xvi, 487 pp., 1997.

122: Timothy G. Gregoire, David R. Brillinger, Peter J. Diggle, Estelle Russek-Cohen, William G. Warren, and Russell D. Wolfinger (Editors), Modeling Longitudinal and Spatially Correlated Data. x, 402 pp., 1997.

123: D. Y. Lin and T. R. Fleming (Editors), Proceedings of the First Seattle Symposium in Biostatistics: Survival Analysis. xiii, 308 pp., 1997.

124: Christine H. Müller, Robust Planning and Analysis of Experiments. x, 234 pp., 1997.

125: Valerii V. Fedorov and Peter Hackl, Model-Oriented Design of Experiments. viii, 117 pp., 1997.

126: Geert Verbeke and Geert Molenberghs, Linear Mixed Models in Practice: A SAS-Oriented Approach. xiii, 306 pp., 1997.

127: Harald Niederreiter, Peter Hellekalek, Gerhard Larcher, and Peter Zinterhof (Editors), Monte Carlo and Quasi-Monte Carlo Methods 1996. xii, 448 pp., 1997.

128: L. Accardi and C.C. Heyde (Editors), Probability Towards 2000. x, 356 pp., 1998.

129: Wolfgang Härdle, Gerard Kerkyacharian, Dominique Picard, and Alexander Tsybakov, Wavelets, Approximation, and Statistical Applications. xvi, 265 pp., 1998.

130: Bo-Cheng Wei, Exponential Family Nonlinear Models. ix, 240 pp., 1998.

131: Joel L. Horowitz, Semiparametric Methods in Econometrics. ix, 204 pp., 1998.

132: Douglas Nychka, Walter W. Piegorsch, and Lawrence H. Cox (Editors), Case Studies in Environmental Statistics. viii, 200 pp., 1998.

133: Dipak Dey, Peter Müller, and Debajyoti Sinha (Editors), Practical Nonparametric and Semiparametric Bayesian Statistics. xv, 408 pp., 1998.

134: Yu. A. Kutoyants, Statistical Inference For Spatial Poisson Processes. vii, 284 pp., 1998.

135: Christian P. Robert, Discretization and MCMC Convergence Assessment. x, 192 pp., 1998.

136: Gregory C. Reinsel, Raja P. Velu, Multivariate Reduced-Rank Regression. xiii, 272 pp., 1998.

137: V. Seshadri, The Inverse Gaussian Distribution: Statistical Theory and Applications. xii, 360 pp., 1998.

138: Peter Hellekalek and Gerhard Larcher (Editors), Random and Quasi-Random Point Sets. xi, 352 pp., 1998.

139: Roger B. Nelsen, An Introduction to Copulas. xi, 232 pp., 1999.

140: Constantine Gatsonis, Robert E. Kass, Bradley Carlin, Alicia Carriquiry, Andrew Gelman, Isabella Verdinelli, and Mike West (Editors), Case Studies in Bayesian Statistics, Volume IV. xvi, 456 pp., 1999.

141: Peter Müller and Brani Vidakovic (Editors), Bayesian Inference in Wavelet Based Models. xiii, 394 pp., 1999.

142: György Terdik, Bilinear Stochastic Models and Related Problems of Nonlinear Time Series Analysis: A Frequency Domain Approach. xi, 258 pp., 1999.

143: Russell Barton, Graphical Methods for the Design of Experiments. x, 208 pp., 1999.

144: L. Mark Berliner, Douglas Nychka, and Timothy Hoar (Editors), Case Studies in Statistics and the Atmospheric Sciences. x, 208 pp., 2000.

145: James H. Matis and Thomas R. Kiffe, Stochastic Population Models. viii, 220 pp., 2000.

146: Wim Schoutens, Stochastic Processes and Orthogonal Polynomials. xiv, 163 pp., 2000.

147: Jürgen Franke, Wolfgang Härdle, and Gerhard Stahl, Measuring Risk in Complex Stochastic Systems. xvi, 272 pp., 2000.

148: S.E. Ahmed and Nancy Reid, Empirical Bayes and Likelihood Inference. x, 200 pp., 2000.

149: D. Bosq, Linear Processes in Function Spaces: Theory and Applications. xv, 296 pp., 2000.

150: Tadeusz Caliński and Sanpei Kageyama, Block Designs: A Randomization Approach, Volume I: Analysis. ix, 313 pp., 2000.

151: Håkan Andersson and Tom Britton, Stochastic Epidemic Models and Their Statistical Analysis. ix, 152 pp., 2000.

152: David Ríos Insua and Fabrizio Ruggeri, Robust Bayesian Analysis. xiii, 435 pp., 2000.

153: Parimal Mukhopadhyay, Topics in Survey Sampling. x, 303 pp., 2000.

154: Regina Kaiser and Agustín Maravall, Measuring Business Cycles in Economic Time Series. vi, 190 pp., 2000.

155: Leon Willenborg and Ton de Waal, Elements of Statistical Disclosure Control. xvii, 289 pp., 2000.

156: Gordon Willmot and X. Sheldon Lin, Lundberg Approximations for Compound Distributions with Insurance Applications. xi, 272 pp., 2000.

157: Anne Boomsma, Marijtje A.J. van Duijn, and Tom A.B. Snijders (Editors), Essays on Item Response Theory. xv, 448 pp., 2000.

158: Dominique Ladiray and Benoît Quenneville, Seasonal Adjustment with the X-11 Method. xxii, 220 pp., 2001.

159: Marc Moore (Editor), Spatial Statistics: Methodological Aspects and Some Applications. xvi, 282 pp., 2001.

160: Tomasz Rychlik, Projecting Statistical Functionals. viii, 184 pp., 2001.

161: Maarten Jansen, Noise Reduction by Wavelet Thresholding. xxii, 224 pp., 2001.

162: Constantine Gatsonis, Bradley Carlin, Alicia Carriquiry, Andrew Gelman, Robert E. Kass Isabella Verdinelli, and Mike West (Editors), Case Studies in Bayesian Statistics, Volume V. xiv, 448 pp., 2001.

163: Erkki P. Liski, Nripes K. Mandal, Kirti R. Shah, and Bikas K. Sinha, Topics in Optimal Design. xii, 164 pp., 2002.

164: Peter Goos, The Optimal Design of Blocked and Split-Plot Experiments. xiv, 244 pp., 2002.

165: Karl Mosler, Multivariate Dispersion, Central Regions and Depth: The Lift Zonoid Approach. xii, 280 pp., 2002.

166: Hira L. Koul, Weighted Empirical Processes in Dynamic Nonlinear Models, Second Edition. xviii, 440 pp., 2002.

167: Constantine Gatsonis, Alicia Carriquiry, Andrew Gelman, David Higdon, Robert E. Kass, Donna Pauler, and Isabella Verdinelli (Editors), Case Studies in Bayesian Statistics, Volume VI. xiv, 394 pp., 2002.

168: Susanne Rässler, Statistical Matching: A Frequentist Theory, Practical Applications and Alternative Bayesian Approaches. xviii, 238 pp., 2002.